TAIGU GENIC MALE-STERILE WHEAT

Developments in Crop Science

Volume 1   Oil Palm Research, edited by R.H.V. Corley, J.J. Hardon and B.J. Wood

Volume 2   Application of Mutation Breeding Methods in the Improvement of Vegetatively Propagated Crops, by C. Broertjes and A.M. van Harten

Volume 3   Wheat Studies, by H. Kihara

Volume 4   The Biology and Control of Weeds in Sugarcane, by S.Y. Peng

Volume 5   Plant Tissue Culture: Theory and Practice, by S.S. Bhojwani and M.K. Razdan

Volume 6   Trace Elements in Plants, by M.Ya. Shkolnik

Volume 7   Biology of Rice, edited by S. Tsunoda and N. Takahashi

Volume 8   Processes and Control of Plant Senescence, by Y.Y. Leshem, A.H. Halevy and C. Frenkel

Volume 9   Taigu Genic Male-Sterile Wheat, edited by Deng Jingyang

DEVELOPMENTS IN CROP SCIENCE (9)

TAIGU GENIC MALE-STERILE WHEAT

Edited by DENG JINGYANG

Institute of Crop Breeding and Cultivation, Chinese Academy of Agricultural Sciences, Beijing, China

SCIENCE PRESS

Beijing, China

ELSEVIER

Amsterdam — Oxford — New York — Tokyo   **1987**

Responsible Editors   Jiang Boning   Peng Keli

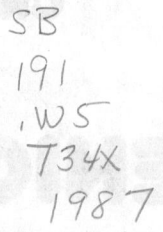

Published by Science Press, Beijing

The Distribution of this book is being handled by the following publishers
      for the U.S.A. and Canada
Elsevier Science Publishing Company, Inc.
52 Vanderbilt Avenue
New York, New York 10017, U.S.A.
      for the People's Republic of China
Science Press, Beijing
      for all remaining areas
Elsevier Science Publishers
P.O. Box 211, 1000 AE Amsterdam, The Netherlands

ISBN 0-444-42644-2 (Vol. 9)
ISBN 0-444-41617-X (Series)
ISBN 7-03-000239-3/Q·44 (外)
Science Press Foreign Language Book No.107

Copyright © 1987 Science Press and Elsevier Science Publishers B.V.

All rights reserved. No part of this publication may be reproduced, stored in retrieval system, or transmitted in any form or by any means, electronic, mechanical, photocopying, recording or otherwise, without the prior written permission of the copyright owner.

Printed by C & C Joint Printing Co., (H.K.) Ltd.

CONTENTS

Foreword
   Lu Liangshu .... (1)
Preface
   Deng Jingyang................................... (3)
The Discovery of the Taigu Genic Male Sterile Wheat
   Gao Zhongli .... (5)
Determination of a Dominant Male Sterile Gene and
   Its Prospect in Practice
   Deng Jingyang & Gao Zhongli..................... (7)
Further Study on the Inheritance
   Ji Fenggao & Deng Jingyang .... (22)
Locus of the *Ta*1 Gene
   Liu Binghua & Deng Jingyang..................... (32)
The Cytological and Histological Investigation
   of the Abortive Process
   Li Xiangyi & Deng Jingyang .... (47)
Submicroscopic Observations on Anther Tissue
   and Microsporogenesis Process I
   Chen Zhu Xizhao .... (61)
The Abnormal Change of Lysine-Rich Histone in
   the Pollen Mother Cell Nuclei Before and
   after Meiosis
   Xue Guangxing & Deng Jingyang .... (74)
Comparative Analysis of Free Amino Acids in
   Sterile and Fertile Anthers
   Zhu Guanglian & T.H.Tsao .... (86)

The Source and Utilization of Free Porline in
   Fertile Anthers and Their Relation to the
   Abortion of Sterile Anthers
   Zhu Guanglian, Sun Chao & T.H.Tsao.............. (97)

Application in Wheat Conventional Breeding and
  Recurrent Selection
  Deng Jingyang & Ji Fenggao .... (110)
Studies on the Combination of Base Population in
  Population Improvement
  Wang Jinxian, Wu Kezhong & Zhou Mingfu .... (119)
Selection and Mating in Recurrent Selection
  Zhang Shaonan & Ji Fenggao .... (130)
Development of Gene Pool with Improved
  Resistance to Scab
  *Wu Zhaosu, Shen Qiuquan, Lu Weizhong
  & Yang Zanlin* .... (134)
A Recurrent Selection Scheme for Breeding
  Wheat Variety with Resistance to Scab
  (Gibberella zeae(Schw.)Petch) Colonization
  Zhang Leqing .... (145)
In the Multiple Resistance Breeding
  Lin Zuoji, Wang Zhenfu & Shuang Zhifu........... (151)
The Recurrent Selection Scheme for
  Improvement of Multiple Characters
  Shen Qiuquan.................................... (158)
As Effective Tool in Distant Hybridization
  Zhu Hanru & Li Xiaochun........................ (162)
Application of the *Ta*1 Gene to Triticale
  Breeding Program
  Ji Fenggao & Deng Jingyang .... (167)

Foreword

Since the discovery and determination of the dominant single gene which controls the Taigu genic male-sterile wheat, much research has been done using it as experimental material in the field of botany, cytology, physiology, biochemistry, genetics, and plant breeding. Although it is not very long since beginning the research, many scientists have already come to believe that this discovery has a significant value for developing wheat genetics and breeding as well as scientific experiments, and for enriching biological theory. In recent years through international academic exchange many biologists have focused their attention upon this discovery. Under the concern of the Chinese government and with the close collaboration of many disciplines, rich fruits have been reaped through the "relay race" of scientific research. This academic monograph is not only a general review of the scientific research on this project, but also a very good tool of communication for all the scientists working on this project.

The major contents are as follows: the discovery and determination of Taigu genic male-sterile wheat, the exploring of new breeding ways, as in recurrent selection by using Taigu genic male-sterile wheat, the development of theories and methods for the establishment of a wheat gene pool, and the research work on the mechanism of sterility for Taigu genic male-sterile wheat, etc. I am fully convinced that the intention expressed and the work done by the authors will surely benefit all the readers. Some practical research ideas are sure to follow. Simultaneously, there exist many arguments waiting for our colleagues to examine and many works waiting for scholars to explore.

I wish each scientist to achieve even greater success by virtue of their diligence.

Prof. Lu Liangshu
August 25, 1985
Beijing

Preface

So far, only a few dominant genic male-sterile mutants were found in crops. The spontaneous mutant first discovered in wheat is the Taigu genic male-sterile wheat, the fertility of which is controlled by a single dominant male-sterile gene. Because of its unique characteristics, it is of very importance in genetics and wheat breeding. Firstly, the sterile plants always give rise to one half sterile progenies and one half fertile ones, the former can set seeds by open-pollination, while the latter can produce seeds by selfing. As all know, open-pollination favors gene recombination, while selfing helps homogeneity and stability. Therefore, the sterile plants, like a gene acceptor, can receive the alien genes extensively and bring about gene recombination continually. The genic recombinants will become homozygous stable plants by selfing of their fertile segregates, and the best ones will be eventually selected as excellent varieties. This characteristic of the Taigu genic male-sterile wheat is very useful in conventional wheat breeding, recurrent selection and exploration of gene-pools. Secondly, its male abortion is thorough and stable. The sterile plants have widely open glumes and remarkably protruded stigmata during flowering stage, so that cross-fertility is high. This does not only increase effectiveness as a breeding tool, but also shows its potential value in utilization of heterosis in wheat. Thirdly, its sterility can be transfered to common wheat varieties and also expressed in extensive wheat genetic background carrying different genomic constitution of wheat-related species, which helps to develop even more extensive application of this dominant genic male-sterile material.

This book covers the following aspects:

1. The discovery of Taigu genic male-sterile wheat and some articles concerning the determination of the dominant gene, such as genetic analysis, cytological observation and location of the gene, etc.

2. The study of the mechanism of male abortion of genic male sterility, such as morphological and anatomical observations, cytochemical and biochemical analyses, etc.

3. Theory and method of the new breeding approaches, such as recurrent selection, establishing and developing the gene-pools, and utilization of the wild germplasms.

We hope that this book, published both in Chinese and in English will interest wheat breeders, biologists, university and college teachers and students. We also hope that this book will arouse exchanges of scientific ideas which will push forward the development of the breeding method of genic male-sterility, so as to open a new area in scientific researches.

Deng Jingyang

THE DISCOVERY OF THE TAIGU GENIC MALE STERILE WHEAT

Gao Zhongli

(*Institute of Wheat Sciences of Taigu County, Shanxi Province*)

In 1971, our research group conducted searching activities of male-sterile wheat plant in Guojia bao, Taigu County. About 10 male sterile plants were found in the field that year but all were self-fertile at harvest of the following year. Having analyzed the reason for failure, we recognized that male sterile plants we found in the previous year grew weak and small, which could have been caused by the environmental conditions. Therefore, we paid greater attention to looking for the male sterile plants different in form from those found last year. We searched through about 500 *mu* wheat production field but none were found. As we were searching in the propagation field for 2-2-3 line, we suddenly saw a vigorous awnless male sterile wheat plant quite distinct from the last year's plants.

In 1973–1974, our research group utilized it in testcrosses, backcrosses and sib mating and found that the percentage of sterile plants varied with the combinations and size of populations; however overall ratio of sterile and fertile plants approached 1:1. In large numbers of bagging test, sterile plants showed 100% sterility and fertile plants set seeds as normal wheat with no segregation occuring in the following generations.

In 1974 winter, Guo Weiyi and I took the material to Beijing and conducted greenhouse studies together with Wang Lianqing, Chen Jinyuan and Shi Jinguo from the Institute of Application of Atomic Energy in CAAS. Then we went to Heilongjiang for accelerating the generations. They thought it was a new type of sterile plant entirely different from the sterile materials already found inside and outside of China.

In 1975–1976, We carried out an experiment in our production brigade in an attempt to form a complete set of 3 lines but failed. In 1977, we visited 14 national institutions but the identification of this material was still unsolved. In respect to the sterile plant, characteristic of no segregation among fertile progenies, we did two things: 1) to search for lines which would maintain the sterility of this sterile plant; 2) to select new strains from the fertile plants. In recent years, we have done more than 600 combinations of testcrosses, backcrosses and sib matings. According to the incomplete calculations, 14,114 plants segregated into 6,885 sterile plants, making up 48.8% of the total; 7,229 fertile plants comprised the remains 51.2%. No stable maintaining line developed out of these combinations. Nevertheless, 30 superior lines have been selected, four of which were entered in the yield test in Taigu County. The best one, 796-15 line, was tested for its regional adoptation in Shanxi Province. Its yield performance ranked first among all varieties, 10.6% higher than that of the control variety Jinmai No.8. It also participates the regional testing.

In February of 1979, Deng Jingyang, from the Institute of Crop Breeding and Cultivation in CAAS, came to our brigade and analyzed all experimental results in comparison with his own studies and concluded that the material is "non-pollen type" sterile plant with a single dominant gene responsible for its male sterility. He also pointed out that it was not appropriate to form a whole set of 3 lines but could be used in many breeding approaches such as open-pollination, directed transfer and recurrent selection. After its identification, this material has received great attention from leaders of all levels, and the 2-2-3 sterile line, once thought useless because the set of 3 lines could not be accomplished, was given a new life and named Taigu male sterile wheat.

DETERMINATION OF A DOMINANT MALE STERILE GENE AND ITS PROSPECT IN PRACTICE

Deng Jingyang

(Institute of Crop Breeding and Cultivation, Chinese Academy of Agricultural Sciences)

Gao Zhongli

(Institute of Wheat Sciences of Taigu County, Shanxi, Province, China)

Introduction

In 1972, an extensive investigation in search for male-sterile wheat plants was carried out in Taigu County, Shanxi Province. Gao Zhongli, a young agro-technician, found a male sterile plant from the cultivated wheat variety in the field[1]. In 1976 the Taigu male sterile wheat was introduced and in 1980 a determinating summation was published[1,2], showing that this male sterile plant belongs to the non-pollen type, that it has a cytoplasm with normal fertility, and that its sterility is controlled by a single dominant male-sterile gene. This mutant was discovered for the first time in wheat and it has been given the name of Taigu genic male-sterile wheat. The dominant male-sterile gene has been designated as $Ta1$[2]. Since 1980, cooperative research of various courses has been organized in different ecological regions in China. The various topics of these studies will be reported respectively. This paper mainly involves the process of the determination of Taigu genic male sterile wheat, the discussion of its value in genetics and wheat breeding, and the presentation of the recent advances of the study by the collaborated research group.

Morphological and Cytological Observations

The original combination of the Taigu genic male sterile wheat is: (Early Maturity 1/30600/2/Taigu 49/Soviet 1/3/Zaoyang/4/Xiaohong/963/5/Taigu 49/San Pastore) F_{10}

Morphological observation

The Taigu male sterile wheat has a loose ear with glumes widely opened and semi-transparent in sunlight, short filaments of poor elongation, thin-shrivelled arrow-like anthers greyish in color, indehiscent, without pollen and no fecundation if bagged. However, the pistil is normal, stigmata protrude remarkably, and allogamous fecundation frequency is high. There is not much difference between the Taigu male sterile wheat and the other self-pollinated varieties in the frequency of seed-set in normal period, i.e. most seeds are borne up to the top of the ear. Some spikelets have 3–4 seeds with milking and germination rate similar to the common varieties. Thus the male sterile wheat, dependent upon allogamy to reproduce, shows absolute sterility and should open many new avenues for wheat breeding (Plate I, II).

The fertile plants segregated from the succeeding generations of the Taigu genic male sterile plant have normal pistils and stamens. No segregation of fertility in the offsprings is observed. These characteristics offer another advantage in the utilization of Taigu genic male sterile wheat (Plate I, 2).

Cytological observation

1. The degeneration period between the open-pollinated plants of Taigu genic male sterile wheat is quite varied; it can take place anytime from the sporocyte to the uninuclear pollen stage, but occurs mostly as the microspores are liberated from their maternal membrane. Whether degeneration is early or late and irrespective of varying morphological appearances, the decomposition is rapid and thorough, only the empty pollen sacs left with just two layers of empty-mast anther wall; so we call it non-pollen type[3]. This characteristic of Taigu genic male sterile wheat is general to its importance as a breeding tool (Plate II, 1).

2. In the anthers of most male sterile plants, more or less of the PMCs form tetrads after reduction division[3], and during the reduction division diakinesis-metaphase, 21 pairs of chromosomes are observed. i.e. $2n = 42$. Meanwhile, in observation of the 42 chromosomes of the root tip squash, chromosomal abnormality cannot be detected. Thus together with the segregation ratio of F_1, we can rule out the possibilities of the male-sterility of Taigu genic male sterile wheat deriving from unpairing or aberration of the chromosomes.

Plate I
1. Left on top, male-sterile ear, with glumes widely opened and visible well-developed pistil.
2. Left below, fertile ear, with normal pollen-shedding dehiscent anthers.
3. Right, a male-sterile plant: arrow points to a male-sterile ear (bagged), no fecundation; unbagged ears, show allogamous fecundation as in common wheat.

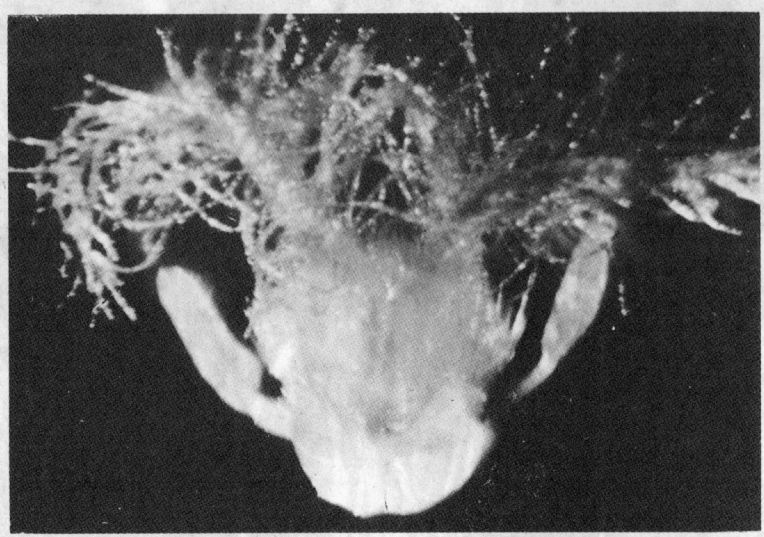

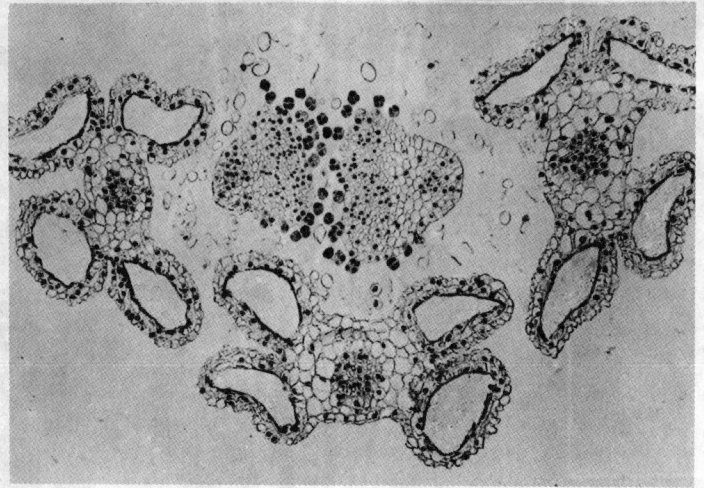

Plate II
1. A male-sterile flower: with small anthers and short filaments of poor elongation; but normal stigmas of possible cross pollination.
2. A male-sterile flower: with twelve pollen-sacs of three anthers, degeneration of all microspores, finally the pollen-sacs are empty only with the residue of the anther-wall.

Genetic Analysis

Analysis of the data of the three Tables 1-3, shows that Taigu genic male sterile wheat has six characteristics:

1. The male sterile plant, when bagged, produces no seeds, fertility of main spike and tillers is the same;
2. If the male sterile plant is crossed with any variety of common wheat, or with the Timopheevi restorer line, or with the fertile plant from the segregation, the fertility of the hybrid F_1 shows a segregation of a ratio of 1 male sterile to 1 fertile;
3. There is no intermediate type of fertility;
4. There is no segregation of fertility in the offsprings from the fertile plant;
5. No modification of fertility can be made even by a number of back-crossings;
6. The characteristic and genetic pattern of the male sterility are not influenced by environmental conditions.

For further illustration on whether or not the fertility of Taigu male sterile wheat was controlled by the nuclear gene, the nucleus and the cytoplasm were analysed separately:

Analysis of the nucleus

According to the characteristics mentioned above, we suppose that Taigu male sterile wheat is a heterozygote, and its male-sterility is controlled by a dominant male sterile gene, we try to probe into it by deduction (Fig. 1).

Tata = dominant male sterile (heterozygote, the supposed Taigu male sterile plant)

tata = fertile (the supposed common wheat)

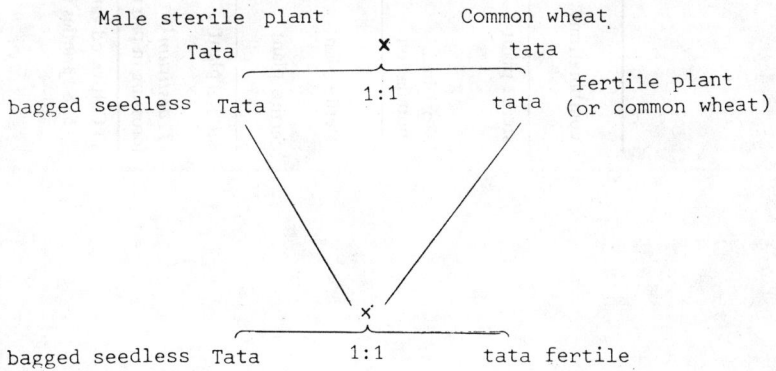

Table 1. Analysis of fertility from segregation of hybrid F_1 of the Taigu male sterile plant (data from Guojiabao)

combinations	years	experimental stations	total of plants	total of sterile plants	total of fertile plants	x^2	P (1:1)
sterile plant × original population	1975	in field	1,497	683	814	11.46	<0.01
	1976	in field	1,252	1,070	1,182	5.570	0.01–0.02
	1976	in green house	763	373	390	0.379	0.50–0.70
	1977	in field	652	319	333	0.301	0.50–0.70
sterile plant × fertile plant	1975	in field	682	310	372	5.636	0.01–0.02
	1976	in field	913	446	467	0.483	0.30–0.50
	1976	in green house	204	101	103	0.0196	0.85–0.90
	1977	in field	36	17	19	0.111	0.70–0.80
sterile plant × T. aestivum[1]	1975	in field	1,634	835	799	0.793	0.30–0.50
	1976	in field	1,543	807	736	3.267	0.05–0.10
(crossing in pair)	1977	in field	3,777	1,851	1,926	1.489	0.20–0.30

1) Cultivated varieties such as Nongda 139, Jinzhong 849 and other lines of subsequent generations, 558 varieties in total.

Table 2. Analysis of segregation of fertility in different combinations or open-pollinated hybrid F_1 under various conditions

years	experimental location	total of combinations or open-pollinated plants	total plants	total of sterile plants	total of fertile plants	x^2	P(1:1)	data source
1975–1977	in field, Taigu	274	11,928	5,978	5,949	0.066	0.80	
1976	in greenhouse, Taigu	126	2,121	1,025	1,096	2.377	0.10–0.20	Guojiabao, Shanxi
1977	in field, Qinghai	145	5,464	2,648	2,816	5.165	0.02–0.05	
1977	in field, Beijing	id.	15	7	8	0.067	0.70–0.80	Research group, Institute of Crop Breeding and Cultivation, Chinese Academy of Agricultural Sciences
1978	id.	5	33	17	16	0.03	0.80–0.90	
1978	id. open-pollination		42	21	21	0	1	
1979	id.	id.	809	396	413	0.357	0.50–0.60	
1979	id. (seeds vernalized)	id.	97	49	48	0.010	0.90–0.95	
1980	in field, Taiyuan	open-pollination	293	147	146	0.003	0.95	Acad. of Agricultural Sciences, Shanxi
1981	in greenhouse, Taigu	8	524	263	261	0.030	0.80–0.90	Guojiabao, Shanxi
1982	in field, Baoding	open-pollination	621	312	309	0.0145	0.90–0.95	Hebei Univ. of Agric.
1982	in field, Jinan	id.	652	327	325	0.006	id.	Acad. of Agric. Sciences, Shandong
1982	in field, Putian	12	762	365	397	1.344	0.20–0.30	Putian Agric. Inst., Fujian

Table 3. Fertility expression in offsprings of fertile plant segregated from hybrid F_1 of Taigu male sterile plant

combinations		years	generations	fertility
selected lines of original population from F_1	G12-18-1	1973–1980	$F_1 - F_8$	all fertile
	G12-18-2	1973–1980	$F_1 - F_8$	all fertile
[(sterile plant × variety A.) × variety B.] × variety C.		1976–1981	$F_1 - F_6$	all fertile
[(sterile plant × variety D.) × variety E.] × variety C. (I)		1976–1981	$F_1 - F_6$	all fertile
[(sterile plant × variety F.) × variety G.] × variety H. (I)		1976–1981	$F_1 - F_6$	all fertile
[(sterile plant × variety D.) × variety E.] × variety C. (II)		1976–1981	$F_1 - F_6$	all fertile
[(sterile plant × variety F.) × variety G.] × variety H. (II)		1976–1981	$F_1 - F_6$	all fertile
30 lines selected from Guojiabao		1974–1983		all fertile 4 lines selected to participate in the provincial region test in 1983

No modification of fertility has been found from 1980–1983 fertile plants selected by 24 breeding units in different ecological regions in China.

The reciprocal crosses of the segregated fertile plant with common wheat were studied. There was no male sterile plant found in both F_1 and its subsequent generations. This indicates that the fertile gene of the fertile plant is homozygous recessive; the fertile plants have only one kind of androgamete, i.e. gamete involving the fertile gene, thus we can eliminate the possibility of the sterility of the gametophyte. We are sure, therefore, that the fertility of the male sterile plant is controlled by a dominant male sterile gene (Fig. 1).

In 1983, Ji Fenggao et al. used rye (*Secale cereale*) to make crosses with Taigu genic male sterile wheat to produce the polyphyletic haploid plants, and then they doubled the chromosomes to make the male sterile gene homozygous artificially. All sterile are the F_1 hybrids of the homozygous male sterile plant, which crosses again with the male fertile plant. The segregation of fertility of the heterozygous F_1 shows a ratio of 1 fertile and 1 sterile, thus once again confirming that the sterility of Taigu male sterile wheat is controlled by a dominant gene. In 1984, by telocentric analysis Liu Binghua et al. found that this gene is located in the short arm of 4D chromosome, and that the centromere distance is 31.1 crossover unit.

Analysis of the cytoplasm

The determination of fertility of the cytoplasm involves great difficulty. Methods of injection or grafting were tried in vain. Consider fertile plant segregated from the descendants of the male sterile plant after crossing, i.e. the cytoplasm originated from the male sterile plant. The offsprings of the fertile plant segregated from the male sterile plant crossed reciprocally with common wheat do not show any male sterile plants. This indicates that the cytoplasmic fertility of Taigu genic male sterile wheat is normal. This cytoplasm of normal fertility and the cytoplasm of the variety Early Maturity I, i.e. the original female parent of the previous hybrid of 6 different lines, are the same. So, the male sterility of the Taigu male sterile plant must be only controlled by the dominant male sterile gene of the nucleus, which is a typical case of genic male sterility.

Analysis of the origin of *Ta*1 gene and its naming

We have excluded various possibilities which tend to induce sterility, such as the unpairing of chromosomes, chromosomic aberration and male sterile gametophyte etc. At the same time an extensive literature on this topic had been reviewed. It was reported that 88 spontaneous male sterile mutants had been found among 48 vegetations, 81 of which were controlled by recessive genes(including 8 of wheat) and only 7 by dominant ones, namely:

Three occasions in cotton (Allison, 1964; Weaver, 1971; Bowman, 1979);

Two in lettuce (Ryder, 1963; 1971);

One in Centranthus (Sobrinho, 1946);

One in potato (Salaman, 1910).

Besides, Franckowiak et al. (1976) and Sassakuma et al. (1978) used EMS treatment for the progenies of a generic hybrid of wheat and obtained a male sterile plant with abortive pollens, F_1 separated into sterile and fertile plants with a ratio of 1:1, the authors presume that it is controlled by a dominant male sterile gene.

The Taigu male sterile plant was discovered in Shanxi in 1972. At that time, tens of thousands of sterile plants were gathered, but only one could perpetuate because its male sterility is controlled by a dominant gene and it is crosscompatible with other common wheat. It is because of this characteristic that Gao Zhongli had the chance to discover it. Spontaneous mutation of homozygote is possible, but it has to go through hybridization to become heterozygote before it can be perpetuated.

Therefore, it is certain that this non-pollen plant with normal fertile cytoplasm controlled by a dominant male sterile gene is a spontaneous mutant, which is discovered for the first time in wheat. This dominant male sterile gene is assigned the name *Ta*1.

Discussion

The discovery of the *Ta*1 dominant male sterile gene opens up a new domain for study of genetics and breeding practice.

In theory

Until now there are a good deal of arguments concerning the theory of male sterility.

Sears (1947), drawing on the previous work in this field claimed that there were 3 types of male sterility, i.e. cytoplasmic, genetic and nucleo-cytoplasmic interaction. His chief reason for making such a conclusion is that, in maize, the T type sterile line could not be restored even though numerous test crosses had been made. Thus the sterility is controlled by the cytoplasm. This is called the trimorphism.

Edwardson et al. (1970), after the restoration of the T type maize, stated that the so-called cytoplasmic male sterility is actually due to the nucleo-cytoplasmic interaction, because the nucleus in the so-called cytoplasmic sterile material is also sterile. Thus the pure cytoplasmic sterility does not exist. This is called the dimorphism. At present, male sterility caused by nucleo-cytoplasmic interaction has given most of 110 types of plants the sterile, maintained and restored lines, but the genic male sterility still lacks convincing evidence.

Kihara et al. (1968) found that a different reaction might result from substitution of cytoplasm and nucleus, i.e. sterile and fertile plants could be obtained accordingly. He suggested that this was a question of nucleo-cytoplasmic disharmony. Each genome or gene has its corresponding cytoplasm. It means that if the gene itself is sterile, its cytoplasm is also sterile. He doubted the existence of the genic male sterility, but actually a maintained line has not yet been found. This is the monomorphism theory.

After a study of so many years, the Taigu genic male sterile wheat cannot find a complete maintained line and a complete restored line, or one can say that it has both; this is an example of typical genic male sterility. Therefore we have good grounds to confirm the theory of dimorphism. Genic male sterility does exist and male sterility due to nucleo-cytoplasmic interaction is a fact. This provides sound scientific evidence relating to this problem in genetics.

In theoretical study, another contribution of Taigu genic male-sterile wheat is that its dominant male sterility is controlled by a single gene which offers a new material for the study of physio-genetics for plants. It is currently in use in several courses for basic theory study, especially in research concerning the mechanism of its sterility. Studies of this topic will be reported separately.

In practice

Self-pollination of crop plants is favorable for maintaining and producing desirable genotypic individuals, but not for recombination of genes; cross-pollination is favorable for recombination of genes, but not for maintaining of desirable genotypic individuals. So Taigu genic male sterile wheat which involves the $Ta1$ male sterile gene possesses both, and will open up several new ways in wheat breeding.

1. Application in traditional breeding

It can be used as a crossing-tool to eliminate emasculation, in simple crossing, complex crossing, step crossing and backcross breeding. It raises the working efficiency and increases the number of combinations which can be made. In complex crossing, the loss of the desirable genes of the primary parents may be reduced. Recently it has been used extensively in different regions in China and rapid advances have been reported. For example, at Guo Jiabao, where the Taigu genic male sterile wheat was discovered, about 30 lines have been selected and some of these lines have participated in the provincial testing. This shows that utilization of Taigu genic male sterile wheat in breeding is a significant improvement over previous methods.

2. Application in recurrent selection

In the seventies of this century, the self-pollinated crop plants as barley, wheat, rice, soybean etc. drew attention gradually as using the recurrent selection method in breeding; but the materials used were recessive male sterile, in which segregation of the progenies is complicated and continues for many generations. Using the Taigu genic male sterile wheat which shows an absolute sterility and stable fertility offers many advantages for recurrent selection of wheat. Research proposals using recurrent selection are being developed for the ecological conditions and breeding objectives of different regions. The organization of populations, control of selection pressure and utilization of the improved populations have been emphasized. The main objective of breeding is resistance to diseases, especially resistance to scab which is poisonous to mankind and animals and other important resistances to rust, powdery mildew (*Erysiphe graminis*), yellow dwarf (byd), root rot (*Ophiobolus graminis*) etc. In adverse conditions, the first thing to think about is drought resistance, saline resistance and so forth. Certainly there are still programmes for increasing yield and quality and all these programmes are being put into effect. From the improved population produced by recurrent selection not only specific varieties can be obtained, but even composite varieties and excellent combined varieties can also be bred. Meanwhile, according to more demanded objectives, some new parental materials can be added before continuing the recurrent selection. The fertile plants selected from recurrent selection can be cultivated in anther culture to get new varieties of vast genetic basis.

3. Application in distant hybridization

In distant hybridization, the difficulties encountered are low frequency of seed-set and the possibility of breeding false hybrids. If the *Ta*1 gene is used as a crossing-tool, it will be more convenient and effective.

In our triticale breeding experiment, we use the good mating varieties having the *Ta*1 gene as bridge varieties in distant crossing. During crossing season, it is possible for one person to obtain thousands of true hybrid seeds. In 1983, Ji Fenggao for the first time succeeded in obtaining the dominant genic male sterile octoploid triticale. This will accelerate the study of chromosome engineering.

4. Application in gene-pool establishing and exploiting

The combined population obtained from the interchange of rich and various genes can be looked upon as a living species gene-pool and can offer new parental materials for wheat breeding. The improved population of different species and generic germplasm produced by distant crossing will enrich the stock of wheat genes and enlarge the genetic mutations of wheat which will enhance the resistance to diseases and adverse conditions. This is a strategy to create heterologous gene-pool artificially.

5. Application of heterosis and character attached marker

Using F_1 heterosis can greatly increase the yield per unit area. At present, the *timopheevi* lines and emasculation by chemical treatment do not give successful results.

The discovery of Taigu genic male sterile wheat controlled by a single dominant gene offers the possibility of producing hybrid seeds by using the two lines method. The basic problem of using the two lines method to produce hybrid seeds requires a character attached marker, which is best differentiated by the color of the seeds, so that sterility and fertility can be divided at the stage of seed-forming and utilization of heterosis of the fertile plants in production can be realized. The Taigu genic male sterile wheat having the character attached marker can be used more conveniently and more effectively in traditional breeding, in recurrent selection, in distant hybridization and especially in basic theoretical studies. At present, research on exploring the character attached marker is underway. Besides, research on using *Ta*1 gene to compose intercrossing populations for selection of test varieties with general combining ability is also having good results.

Taigu genic male sterile wheat controlled by a single dominant gene acts as a machine-tool which makes breeding methods much more efficient. As wheat is one of the staple food crops of mankind, this breeding tool can raise the production and enhance the quality; it has a very important significance in opening up new ways for worldwide agricultural breeding.

References

[1] 高忠丽，1979，攀登小麦育种科研新高峰，山西农业科学，（1）：13—15。
[2] 邓景扬、高忠丽，1980，小麦显性雄性不育基因的发现与利用—太谷不育小麦鉴定总结，作物学报，（2）：85—98。
[3] 李祥义、邓景扬，1983，太谷核不育小麦雄性败育过程的细胞形态学研究，作物学报，（3）：151—156。
[4] 纪凤高、邓景扬，1984，太谷核不育小麦遗传鉴定结果再验证及显性雄性不育八倍体小黑麦的人工合成，中国科学（B辑），（12）：1111—1125。
[5] 刘秉华、邓景扬，1984，太谷核不育小麦显性不育单基因初步定位，山西农业科学，（5）：10—11。
[6] Dang Kien-Duong (Deng Jingyang), Gao Zhongli, 1980, Détermination et utilisation d'un gène dominant mâlestérile chez le blé Taigou (*T. aestivum* L.), *Saussurea,* (11): 27-41.
[7] Deng Jingyang (Dang Kien-Duong), Gao Zhongli, 1982, Discovery and determination of a dominant male sterile gene and its importance in genetics and wheat breeding, *Scientia Sinica* (Series B), (5): 508-520.
[8] Athwal, D. S., P. S. Phul, and J. L. Minocha, 1967, Genetic male sterility in wheat, *Euphytica,* **16**: 354-360.
[9] Briggle, L. W. A., 1970, Recessive gene for male sterility in hexaploid wheat, *Crop Sci.,* **10**: 693-696.
[10] Driscoll, C. J., 1977, Registration of Cornerstone male sterile wheat germplasm, *Crop Sci.,* **17**: 190.
[11] Favret, E. A. and G. S. Ryan, 1966, Mutation in plant Breeding, 49-60.
[12] Fossati A. and M. Ingold, 1970, A male sterile mutant in *T. aestivum, Wheat Inf. Serv.,* **30**: 8-10.
[13] Jan C. C. and C. O. Qualset, 1977, Genetic malesterility in wheat inheritance, *Crop Sci.,* **17**: 791-794.
[14] Pugsley, A. T. and R. N. Oram, 1959, Genic male sterility in wheat, *Austral. Plant Breed. Genet. Newsletter,* **14**.
[15] Крулнов, B. A., 1968, Genic male sterility in common wheat (*T. aestivum* L.), *Soviet Genetics,* **4**(10): 28-35.
[16] Allison, D. C. and W. D. Fisher, 1964, A dominant gene for male sterility in upland cotton, *Crop Sci.,* **4**: 548-549.
[17] Weaver, J. B. Jr. and T. Ashley, 1971, Analysis of a dominant gene for male-sterility in upland cotton, *Gossypium hisutum* L., *Crop Sci.,* **11**: 596-598.
[18] Bowman, D. T. and J. B. Weaver, 1979, Analysis of a dominant male-sterile character in upland cotton, *Crop Sci.,* **19**: 628-630.
[19] Ryder, E. J., 1963, An epistatically controlled pollen sterile in lettuce (*Lactuca sativa* L.), *Proc. Amer. Soc. Hort. Sci.,* **83**: 585-589.
[20] ————, 1971, Genetic studies in lettuce (*Lactuca sativa* L.), *J. Amer. Soc. Hort. Sci.,* **96**:826-828.
[21] Sobrinho, L. G., 1946, Sur le comportement génétique de Centranthus ruber, *D. C. Portugaliae Acta Biol. Ser. A.,* **1**: 385-390.
[22] Salaman, R. N., 1910, Male sterility in potatoes: a dominant Mendelian character, *J. Linn. Soc.* (London), **39**:301-312.

[23] Franckowiak, J. D., S. S. Maan, and N. D. Williams, 1976, A proposal for hybrid wheat utilizing *Aegilops squarrosa* L. cytoplasm, *Crop Sci.,* **16**: 725-728.
[24] Sasakuma, T., S. S. Maan, and N. D. Williams, 1978, EMS induced male-sterile mutants in euplasmic and alloplasmic common wheat, *Crop Sci.* **18**: 850-853.
[25] Sears, E. R., 1947, U. S. Dept. Agr. Yearbook, 245-255.
[26] Edwardson, J. R., 1970, *Bot. Rev.,* 341-401.
[27] Kihara, H., 1968, Cytoplasmic relationships in the Triticinae, 3 Int. Wheat Gen. Symp., 125-134.

FURTHER STUDY ON THE INHERITANCE

Ji Fenggao   Deng Jingyang

(*Institute of Crop Breeding and Cultivation, Chinese Academy of Agricultural Sciences*)

Introduction

In 1972, Gao Zhongli discovered a male sterile plant among the common wheat in Taigu county of Shanxi Province[1]. Several institutions in China made genetic studies of the offsprings of this plant. Since the great achievements of utilizing heterosis in plant breeding, the first thought was to complete the whole set of 3 lines to make hybrid wheat from the new male sterile plant. Nevertheless, F_1 hybrids of this plant always segregate into sterile and fertile plants; no variety or strain was found which can completely maintain or restore its fertility. Deng Jingyang analyzed other's research informations as well as his own experimental results and concluded that the male sterility of the plants was controlled by a single dominant gene[2]. Because it was found in Taigu county, it was named after the county of Taigu and the dominant mono-gene was referred to as *Ta*1.

Dominant male sterile mutations in plants rarely occure; the great majority of mutations being controlled by recessive genes. There is no precedent for a naturally occuring dominant male sterile mutation in common wheat. After Deng's study was published, there were still many contradictory views concerning the genic mechanism involved. Therefore, it is necessary to study the identification of the male sterile wheat further and with different approaches. Since different kinds of the male sterile materials have distinct genetic characteristics and can be used in different way in plant breeding. It is clear that further study, however it relates to the original conclusion, will benefit the research and application of Taigu genic male sterile wheat.

Experimental Design

It is characteristic of the dominant gene that all F_1 hybrids of homozygous dominant plants crossed with the corresponding homozygous recessive plants will show the dominant character while the F_1 hybrids between the heterozygous dominants and homozygous recessives will segregate into 1:1 ratio. According to the original identification, *Ta*1 gene can only exists in a monogenic state because the plant carrying *Ta*1 gene can not be self-pollinating. On the basis of Mendelian genetics, male sterility and male fertility belong to a pair of relative traits generally controlled by alleles. Therefore, we suppose that the allele of *Ta*1 gene is *ta*1, so that common wheat and fertile plants from the sterile ones are not segregated in fertility and their genotype should be *ta*1*ta*1. F_1 hybrids of sterile plants are always segregated into half fertile and half sterile ones; their genotype should be *Ta*1*ta*1. Because the sterile plants are heterozygous, they should produce two kinds of female gametes: *ta*1 and *Ta*1. If these female gametes can be parthenogenetically developed into mature plants, the phenotype of the stamen of these plants might represent the genotype of those female gametes. Then by doubling the chromosome number of those haploid plants, *Ta*1 and *ta*1 genes will be artificially homogenized into *Ta*1*Ta*1 and *ta*1*ta*1 respectively. The F_1 hybrids of the homozygous sterile plants should all carry *Ta*1 gene and hence all show male sterile.

In the beginning we tried 3 approaches: 1) To pollinate the male sterile Taigu wheat with *Hordeum bulbosum* pollen to get haploid young embryoes which have only the genome of the female parent because of the elimination of the chromosomes of the male parent[4]; 2) To treat the stigmas of the sterile plants with chemicals to induce parthenogenesis; 3) To pollinate the male sterile plants with rye pollen to obtain polyhaploid plants. In the first two ways, haploids with pure common wheat genome could be obtained and there was no obstacle for the *Ta*1 gene expression. However, as we found that rye genome has no clear effects on the expression of *Ta*1 gene, the experiment became mainly focused on the 3rd way. The results of the experiment in this respect will be mainly reported in this paper.

The genomic constitution of common wheat is AABBDD. That of diploid rye is RR. F_1 generation between Taigu male sterile wheat and rye should have two kinds of genotype: ABDR(*Ta*1) and ABDR(*ta*1). After chromosome doubling, they become AABBDDRR (*Ta*1*Ta*1) and AABBDDRR(*ta*1*ta*1). The latter is the newly bred octaploid primary triticale strain. The former is the homozygous male sterile octaploid triticale. Pollinating with the normal octaploid triticale, F_1 hybrid should all have $Ta1$ gene, i.e. all show the male sterile.

Materials and Methods

Six female parents carrying $Ta1$ gene were derived from the sterile plants of F_1 hybrids of open-pollinated Taigu wheat crossed with spring wheat PH88, (Zhong 7605/PH88)F_5, Penjamo 62, Zhong 7725, Yecora and Zhong 7901, which were kindly supplied by the spring wheat breeding lab of our institute. The progenies of hybrids and backcrosses were produced by authors. The male parents were Jing Zhou rye, Snoopy – 2, Snoopy – 3 and Lan Zhou rye, which were given by polyhaploid breeding lab of our institute. Parents were identified as hexaploid wheat with 42 chromosomes (Plate III, 1) and diploid rye with 14 chromosomes (Plate III, 2) by the cytological study of PMCs fixed in Carnoy mixture (3:1) and squashed in aceto carmine.

The chromosome number of F_1 hybrids were doubled by submerging the tillering nodes of young plants in the solution of colchicine and dimethyl sulphoxide.

The treated sterile plants were pollinated individually with the pollen of triticale plot 12 ($2n=8x=56$), plot 20 ($2n=8x=56$) and Jian 45 strain ($2n=6x=42$). Control I was the male sterile wheat carrying $Ta1$ gene. Control II was the F_1 haploid plants without $Ta1$ gene from the cross between Chinese Spring wheat and rye, some of which was treated with colchicine.

The experiment was carried out in the green houses of Dong Sheng experimental station and of our institute as well as in the fields.

Results and Analysis

The stamen characteristics of haploid plants and their genotypic analysis

The F_1 plants of all combinations between the genic male sterile wheat and rye can be classified into two distinct groups at flowering stage: the large anther plants and the small anther ones (Plate III, 5–6).

In the large anther plants, like the control II without $Ta1$ gene, the anther appears milk-white in the early developing stage, becomes green after meiosis and then turns to yellow green around the flowering. At anthering stage, the filament can stretch and send anthers out of glumes but anthers which are slender and weak could not dehisce and shed pollen grain. Thus, very few plants set seeds without alien pollination (Plate III,7).

In small anther plants, most of the anthers stop growing before or after meiosis and remain at young milk-white state. At flowering stage, there are only some remains of dead anther wall with no pollen in it. Meantime, the filament is dead too, so that it can not stretch out (Plate III, 8). Several days later, the glumes are loosely open, outside of which no anther can be seen. In a few spring-sowed plants the anther abortion happens a little late but it is not difficult to distinct them from the former plants.

Although both plants are haploid with 28 chromosomes (Plat III, 3-4) and self-sterile, the development of stamen and the morphology of the large anther plants are similar to that of control II without $Ta1$ gene. On the contrary, the development of stamen and morphology of small anther plants are similar to that of control I carrying $Ta1$ gene. The large anther plants probably can not set grains by self-pollination because they are haploid and have abnormal meiosis. The stamen abortion of the small anther plants is caused by the existance of $Ta1$ gene. the phenotypes of these two plants represent the genotype of female gametes produced from Taigu male sterile wheat. The segregation ratio of these two plants can be seen in Table 1.

The trait of stamen and the genotype analysis of the amphipolyploid from the chromosome doubling of the polyhaploid plants

In spring and autumn of 1982, the F_1 seedlings of Taigu wheat / rye were treated with colchicine and the treated plants grew extremely uneven but all plants could be classified into 2 groups of large anther and small anther plants.

The development and morphology of stamen of large anther plants was similar to that of untreated large anther plants. Part of the plants had some well-developed anthers and set seeds by self-pollination, which showed the success of chromosome doubling. Although not all anthers in all plants could dehisce and shed pollen, yellow anthers could be seen hanging out of glumes at flowering.

The development and morphology of stamen of the small anther plants were also similar to those of untreated small anther plants. After being pollinated with pollen of the normal octaploid or hexoploid triticale, the majority of plants could give seeds. 15 out of 17 small anther plants treated in the spring gave some seeds. Not all autumn treated plants were pollinated. After chromosome doubling, the number of large anther and of small anther plants are listed in Table 2.

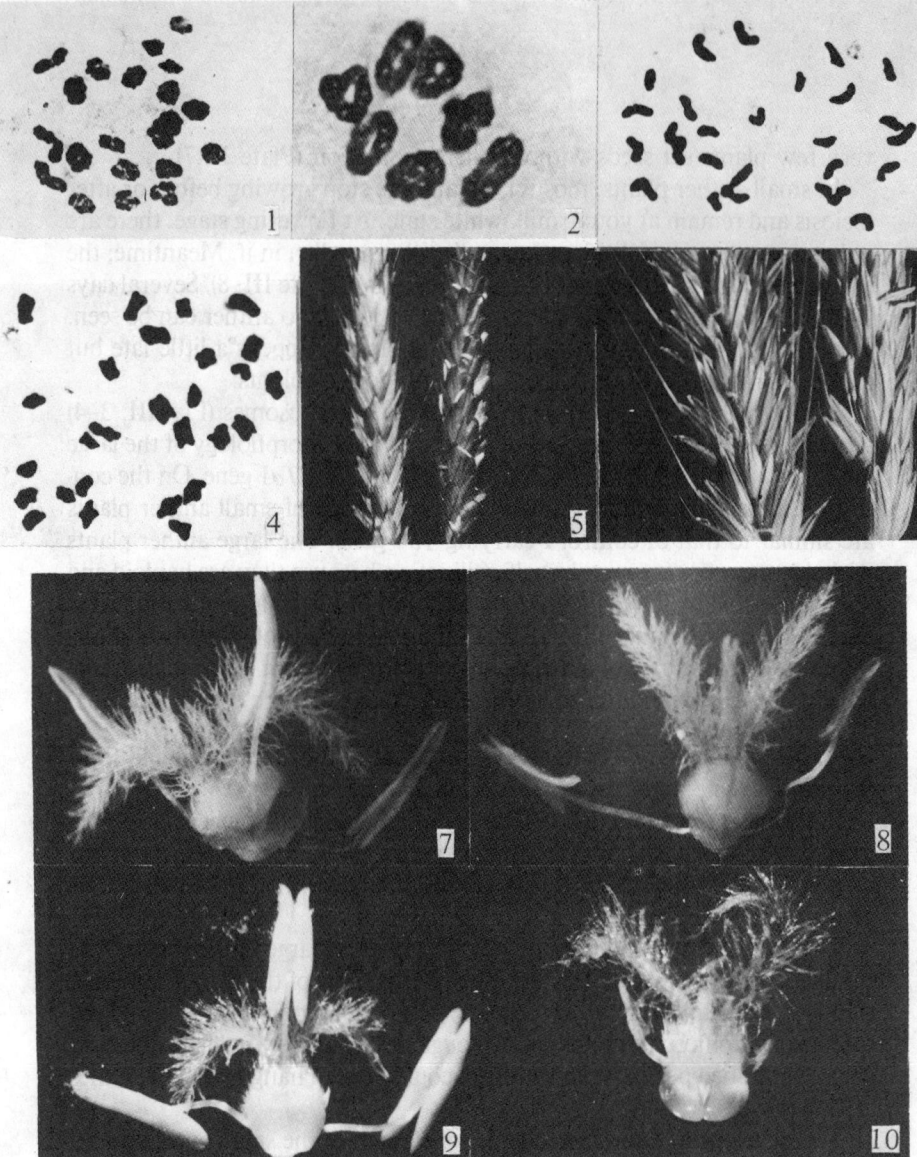

Plate III

1. The Taigu male-sterile wheat with 21 bivalents.
2. Female parent rye with 7 bivalents.
3,4. The large anther plant (3), and small anther plant (4) from the F_1 hybrids between the Taigu male-sterile wheat and rye, all having 28 chromosomes.
5. The appearance of spikes from large anther plant (right) and small anther plant (left) in F_1 hybrid of Taigu male-sterile wheat × rye.
6. The same as 5, the slender anthers can be seen out of glumes in large anther plant (right) but no anther can be observed out of glumes in small anther plant (left).
7. Pistil and stamen of large anther haploid plants.
8. Pistil and stamen of small anther haploid plants.
9. Pistil and stamen of fertile plant from male-sterile octaploid triticale.
10. Pistil and stamen of sterile plant from male-sterile octaploid triticale.

Table 1. Segregation ratio of large anther and small anther plants in the cross of Taigu wheat with rye

time and location of sowing	No. of combinations	No. of total plants	large anther plants	small anther plants	χ^2	P (1:1)
in autumn of 1982	22	595	296	299	0.015	>0.90
in autumn of 1982 transplanting in greenhouse	24	134	68	66	0.030	0.80–0.90
in spring of 1983 in the field	20	280	149	131	1.157	0.20–0.30

Table 2. Segregation ratio of the large anther and small anther plants from the haploid plants treated with colchicine

time and location of sowing	total heading plants	large anther plants	small anther plants	χ^2	P (1:1)
spring of 1982 in the field	27	10	17	1.815	0.10–0.20
winter of 1982 in greenhouse	122	66	56	0.819	0.30–0.50
total	149	76	73	0.060	0.80–0.90

The small anther plants are considered to contain $Ta1$ gene before chromosome doubling. After treatment, part of the meristem in some plants doubled their chromosome number so that $Ta1$ gene was homologized as $Ta1Ta1$. this kind of florets produced at meiosis the female gametes containing $Ta1$ gene. The megasporocyte which failed in chromosome doubling might form at meiosis the unreduced gametes carrying $Ta1$ gene. No matter what happened, the genotype of normal female gametes from the small anther plants should remain the same as those of somatic cell of haploid plants; i.e. all contain $Ta1$ gene. If the megasporocyte of haploid plants produced at meiosis the viable female gametes with 28 or less than 28 chromosomes, except for those lacking the right chromosome carrying $Ta1$ gene, most of them should also have $Ta1$ gene.

The fertility of the offspring of homozygous fertile (self-pollinated) and sterile (artificialy pollinated) plants.

40 seeds were obtained from the doubled fertile plants in the spring of 1982 and planted in a green house in the autumn of the same year; 35 of these grew into full plants. All belonged to the large anther plants. The anther in most of these plants could dehisce and shed pollen grains but their self-fertility varied.

143 grains were produced by the sterile plants (by artificial pollination) and 103 seeds grew into full plants after the autumn sowing. One of these had large anthers, a few of which dehisce and shed pollen, but failed to set grains by self-pollination. Perhaps it lost the chromosome carrying $Ta1$ gene or there were some other reasons. 102 plants were male sterile with the same degree of stamen abortion as that of control I carrying $Ta1$ gene. Being pollinated by octoploid triticale, almost all plants could produce seeds.

The fertility segregation of the progenies of sterile heterozygous plants

In the spring of 1983, 715 early mature grains were harvested from the sterile plants and sown in pots in greenhouses after treatment with hydrogen peroxide solution to break their dormancy. The fertility were checked at flowering stage. Most plants could be clearly distinguished as either fertile or sterile. The anthers of fertile plants were plump and able to shed pollen grains and set seeds by self-pollination (Plate III, 9), while stamens of sterile plants were small. In plants like the Taigu wheat control I, stamen abortion occurred at meiosis or a little later (Plate III, 10) but a few sterile plants had later stamen abortion. No small anther plant has been found to be self fertile. The segregation ratio of sterile plants and fertile plants are shown in Table 3.

Table 3. The segregation ratio of fertile and sterile plants F_1 of the heterozygous male sterile triticale

combination	No. of combination	total plants	fertile plants	sterile plants	x^2	P(1:1)
male sterile octoploid triticale/octoploid triticale	17	445	222	223	0.002	>0.90
male sterile octoploid triticale/hexaploid triticale	6	106	51	55	0.151	0.60–0.70
total		551	273	278	0.045	0.80–0.90

The sterile plants listed in Table 3 were crossed with 126 strains and varieties of 6x or 8x triticale, to utilize male sterile gene in the triticale breeding program.

Conclusion and Discussion

All results are shown in following figure

In general, the results obtained correspond with the inference of the experimental design, therefore the male sterility of Taigu wheat is actually determined by the dominant monogene $Ta1$. Any genic change related to more than one locus or chromosomal structure change will be expressed on the phenotype of the haploid plants or homozygous amphiploid plants as well as the segregation ratio of the sterile and fertile plants.

The offsprings of all segregating generations, haploid or amphiploid, highly fit theoretical ratio, which demonstrate that $Ta1$ gene has no effect on the viability and fertility of female gametes and no influence on the selective fertilization and the viability of the carrier plants. If there were these effects, the segregating ratios of large anther plants and small anther plants

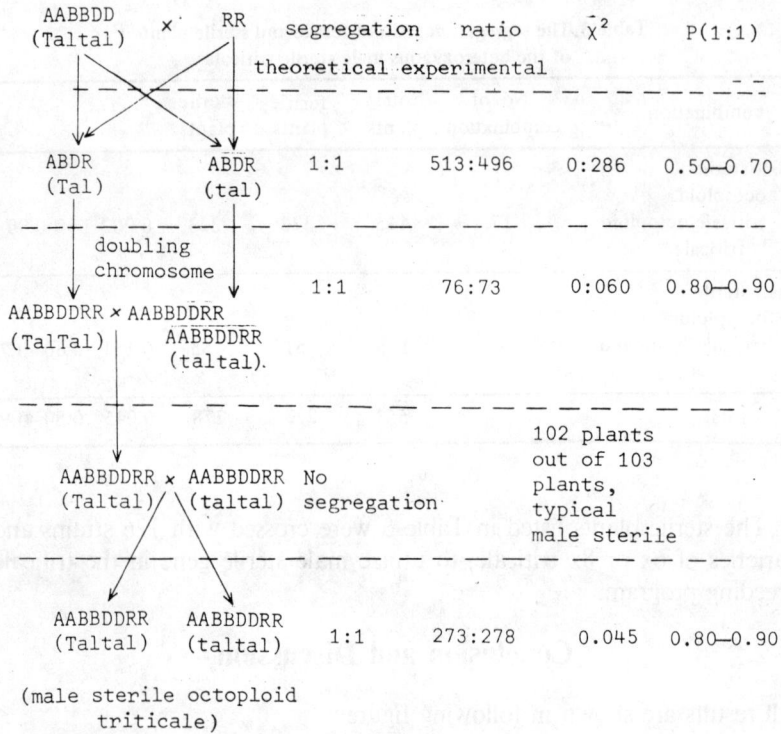

(male sterile octoploid triticale)

in haploid stage would not fit the theoretical ratio 1:1. At present, great attention has been paid to the utilization of genic male sterile gene as a tool for the gene recombination in self-pollinating crops in the world[5]. *Ta*1 gene having none of the above-mentioned harmful effects has out-standing advantages as a tool for gene recombinations. Breeders can use the sterile plants, with their high cross-fertility, for gene recombinations and again use fertile plants segregated from the offsprings of sterile plants to breed new pure strains with superior genotypes or to develope synthetic composites with wide adaptability and disease resistance. Thus, breeders may take the advantage of both self- and cross-pollinating crops to advance wheat breeding[3].

The implication of the *ta*1 gene

We supposed that the allele of *Ta*1 gene is *ta*1, for the convenience of research and description. Now, the results show *Ta*1 gene is the determining gene for male sterility which is a qualitative character; therefore, the gene can be located on a chromosome. So, what is the implication of the *ta*1 gene?

In the synthetic male sterile octoploid triticales or those under transfering, male fertile gene from rye has not been expressed because of the existance of *Ta*1 gene. Common wheat came from 3 related species through evolution. There is no doubt that these 3 related species were themselves normally fertile, containing fertile genes. Thus, common wheat should have 6 genes or 6 groups of genes concerning male fertility while octoploid triticale should have 8 of these genes. Nevertheless, *Ta*1 gene can express itself stably not only in common wheat genome but also in octoploid triticale. It is unknown whether *Ta*1 gene is a mutant of one of fertility genes in common wheat or comes from some other locus. So *ta*1 gene, the allele of *Ta*1 gene, represents the locus allelic to *Ta*1 gene or all fertile genes which have not been expressed in the sterile plants. All fertile genes are considered as "alleles" of sterile gene or as background genes; all these remain to be studied and discussed. In terms of trait analysis, it is more reasonable that *ta*1 gene should be the theoretical representation of several repetitive male fertile genes which have same function and occupy different loci but have no additive effects. It is not certain whether the locus allelic to *Ta*1 is related to the male fertile.

References

[1] 高忠丽，1981，太谷核不育小麦的发现过程，农业科技通讯，（1）：4－5。
[2] 邓景扬、高忠丽，1980，小麦显性雄性不育基因的发现与利用，作物学报，6（2）：85－98。
[3] 邓景扬、纪凤高，1983，显性雄性不育核基因Tal在小麦育种上的利用价值与主要利用途径，中国农业科学，（4）：6－11。
[4] Snape, J. W. and E. Simpson, 1978, the Crossabilities of Wheat Varieties With *Hordeum bulbosum, Heredity,* **42**(3): 291-298.
[5] Suneson, C. A., 1962, Use of Pugsley's Sterile Wheat in Crop Breeding, *Crop Sci.,* 2: 534-535.

Locus of *Ta*1 Gene

Liu Binghua   Deng Jingyang

(*Institute of Crop Breeding and Cultivation, Chinese Academy of Agricultural Sciences*)

Introduction

The Taigu male sterile wheat discovered in China is a common wheat (*Triticum aestivum* L.), the male sterility of which is controlled by a single dominant gene (designated as *Ta*1). Because of its normal cytoplasm, wide opening of glumes and protrusion of stigma at flowering, which is favourable to receive the foreign pollen grains, it is of importance in wheat breeding and genetic study[1]. In order to utilize the sterile gene effectively, it is necessary to ascertain the genome, the chromosome and arm where the gene is located as well as its position in relation to the centromere.

Based on the fertility performance of nullisomic series, Sears made certain that the male fertility genes were on the chromosomes 4A, 4B, 5A, 5B and 5D, then, with the help of the ditelocentric analysis, he further, located the fertility genes on the long arm of these chromosomes[3,4]. Driscoll et al., applied radiation treatment to the monosomics of above-mentioned chromosomes and induced a male sterile mutation on the long arm of chromosome 4A[5,6]. Tsunewaki located the male sterile gene on the chromosome 1B according to the allelic relationship between sterile gene and the restoration gene in speltoid wheat[6]. In 1966, Maan bred a monosomic series containing the sterile cytoplasm (*T. Timopheevi* wheat cytoplasm) in an attempt to locate the genic male sterile gene in the nucleoplasmic interaction type sterile line. No further information on this issue has been reported[7]. We employed a series of new procedures such as genome analysis, telosomic mapping and analysis to locate the dominant male sterile monogene in Taigu wheat. In this paper, we summarize this study.

The Principles of *Ta*1 Gene Location

*Ta*1 gene location in genome

At late meiosis, bivalents will equally split and seperate to two poles, but univalents will randomly move to one of them. A large part of them lag behind, can not reach the pole at telophase I, and eventually disappear in the cytoplasm[2,10]. The Taigu male sterile wheat (AABBDD) is crossed and backcrossed with a recurrent parent — a related wheat species, which shares one or two genomes with common wheat. One or two genomes of Taigu wheat will become univalents at meiosis in the progenies of crosses and backcrosses. The univalents will be lost gradually in the progenies as the backcross generations proceed. If *Ta*1 gene is located on a chromosome of the genome existing in univalent state, the dominant *Ta*1 gene will be lost as the univalent dissapears. Therefore, the male sterile plants carrying *Ta*1 gene in the progenies will be obviously less than the fertile or partial fertile plants without *Ta*1 gene. On the contrary, if the *Ta*1 gene is on chromosome of a genome which formed bivalents, the segregation ratio of sterile plants and fertile (or partial fertile) plants will remain a 1:1 ratio. Thus the genome carrying *Ta*1 gene can be determined.

Telosomic analysis

The principle of telosomic analysis is illustrated in following Fig. 1-3. Taigu wheat with normal chromosome constitution (21″) is crossed with ditelosomic lines (20″ + t″, Plate IV, 1). The offspring will be half sterile and half fertile plants, all of which have heteromorphic disomics (20″ + 1t″, Plate IV, 2). The sterile plants from each telosomic combination are pollinated with the pollen grains from the normal disomics to produce BC_1 plants, in which the half sterile and half fertile plants will be segregated from both critical family line, carrying *Ta*1 gene, and the non-critical family line, without *Ta*1 gene, by the fertility performance. So the critical family line can not be distinguished from the non-critical ones. But their different chromosome constitution can be found by cytological observation. In the non-critical family line, there are both normal disomics and heteromorphic disomics in the sterile or fertile plants. Both are equal in number (Fig. 1). In the critical family line, there are two situations: firstly, if the right chromosome arm is involved, fertile plants are heteromorphic disomic but sterile plant are normal disomics. However, as crossingovers in the

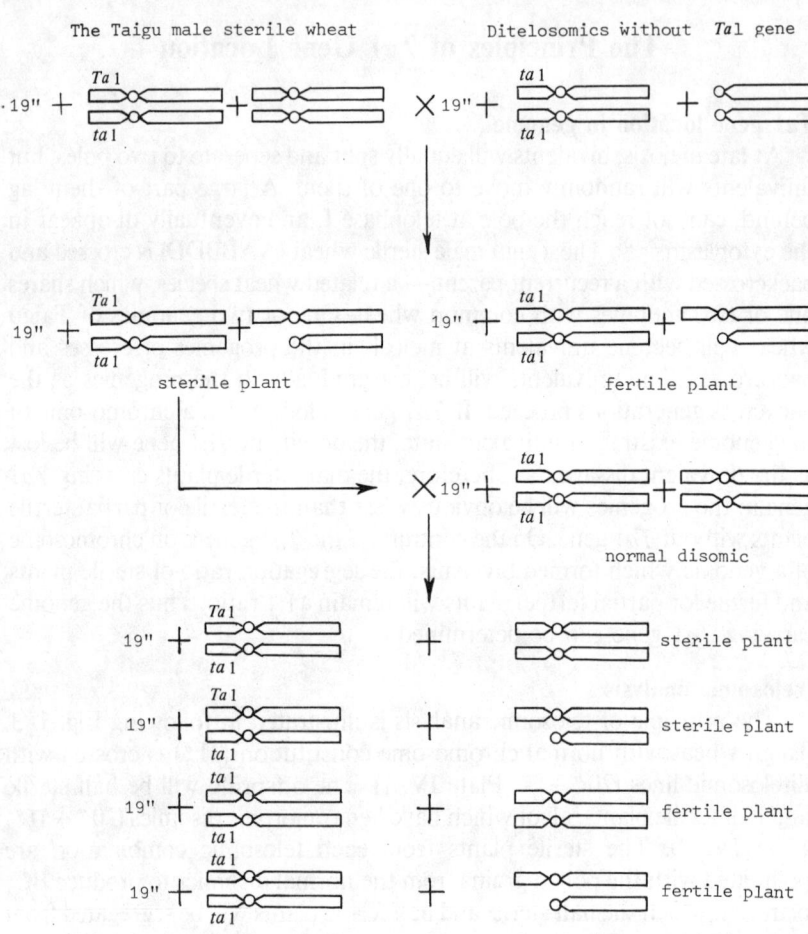

heteromorphic bivalents are possible, some fertile normal disomics and sterile hetermorphic disomics may occur (Fig. 2). Based on the number of the recombinants, the genetic distance between sterile gene and the centromere can be estimated[9,11]. Secondly, in the telocentric family line, which have the opposite arm of the chromosome carrying *Ta*1 gene, fertile plants can only be heteromorphic disomics, and sterile plants normal disomics (Fig. 3). Some sterile telosomics can be verified be double telosomics (20"+t"L+t"s) or nullitetrasomics instead.

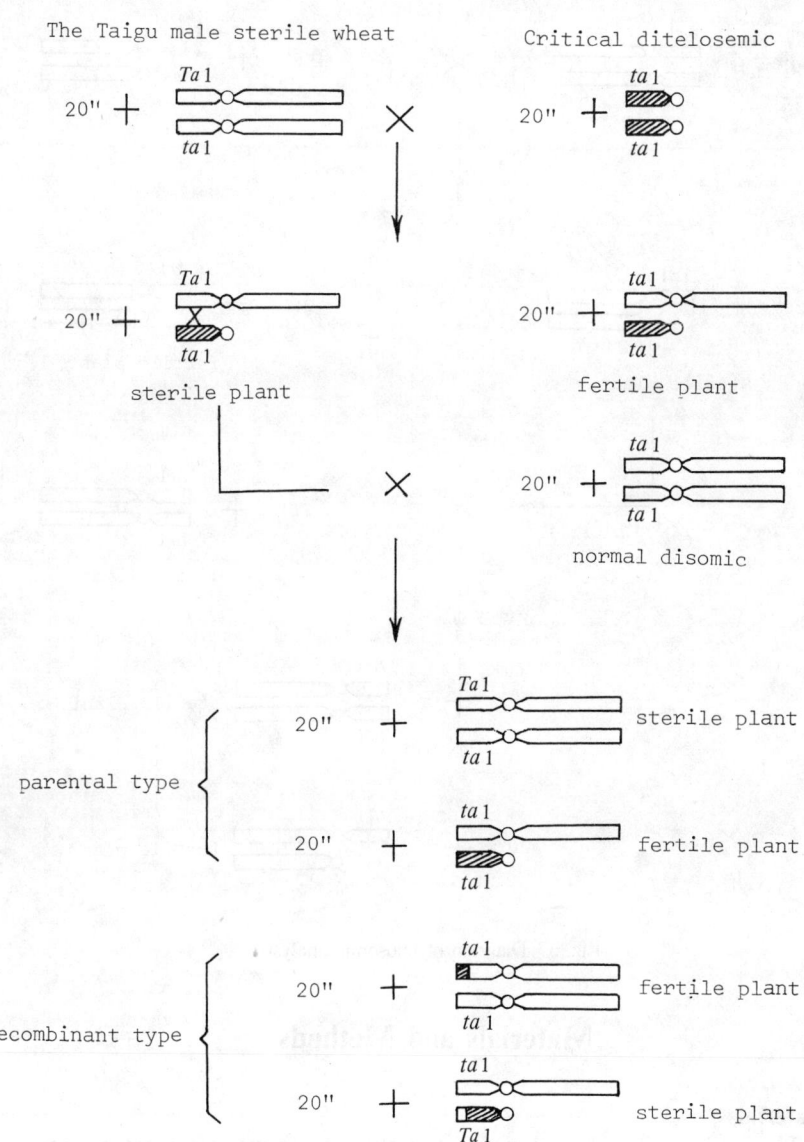

Fig. 2 Diagram of telosomic analysis

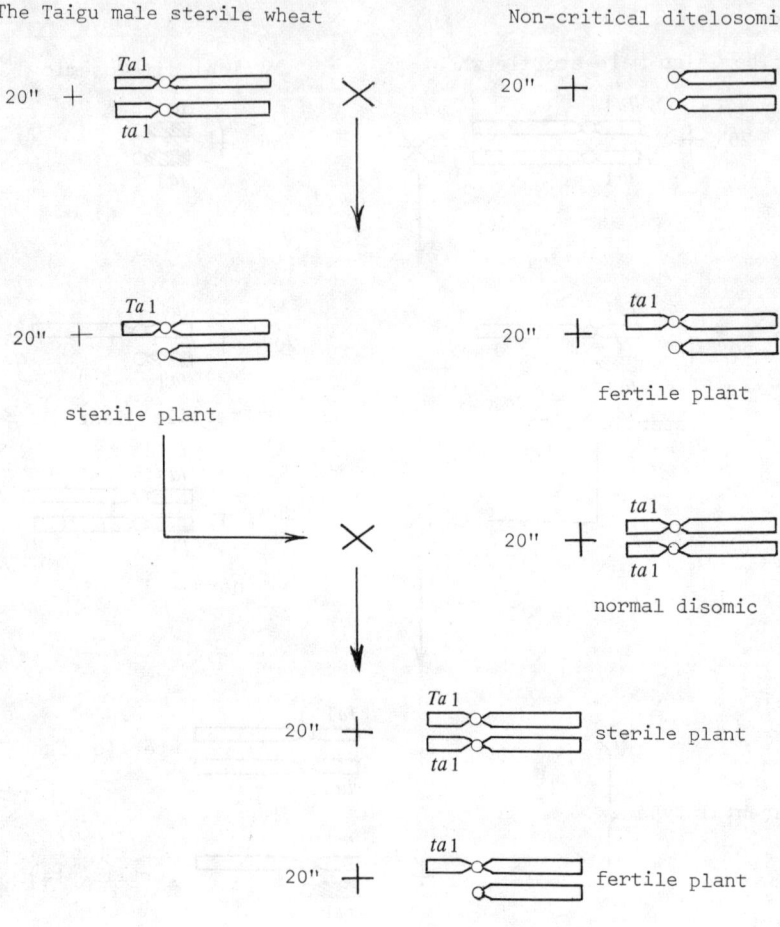

Fig. 3 Diagram of telosomic analysis

Materials and Methods

Materials

The sterile material employed in this study were the offsprings of open-pollinated Taigu male sterile wheat (*T. aestivum*, AABBDD, 2n = 6x = 42). Tetraploid durum wheat Mexicali -75 (*T. durum*, AABB, 2n = 4x = 28) was utilized as male recurrent parent for genome study. Ditelosomics $(20''+t'')$ were used for the telosomic analysis (including telosomic test). Nonda 139 and Chinese Spring were normal disomics used as the backcrossing male parents.

Fertility investigation

Stamens of the Taigu male sterile wheat carrying $Ta1$ gene degenerate into yellow-white, small arrowhead-shaped anthers. Whereas, among the interspecific hybrids, the partial fertile plants, due to the aneuploidy of their pollen, have large anthers with regular yellow colour. Therefore, the sterile plants and the partial fertile (or fertile) plants are distinguishable by the morphology of anthers. The fertility of the hybrids or backcross progenies between durum and Taigu wheat were individually investigated and calculated in every generation. The offsprings of the cross between F_1 sterile plants (Taigu wheat ditelosomics) and the normal disomics were checked for the fertility of each plant in different family lines.

Cytological observation

The appropriate young spikes were fixed in Carnoy's solution (95% ethanol: acetic acid = 3:1) for 24 hours, then stored in 70% ethanol under refrigeration. Anthers were squashed in acetocarmine to study meiosis of PMCs. The chromosome constitution of each plants (in both sterile and fertile plants) were recorded and had typical cells photographed.

Analysis of Results

The $Ta1$ gene location in genome

1. Fertility of F_1 hybrid and backcross progenies

The Taigu male sterile wheat (AABBDD) were pollinated with a durum wheat Mexicali-75 (AABB). The sterile plants carrying $Ta1$ gene were selected from the F_1 hybrids and backcrossed with durum wheat. The backcrossing were repeated between the selected sterile offsprings and durum wheat, the recurrent parent. The fertility of F_1, BC_1, BC_2 and BC_3 were investigated and listed in Table 1.

In F_1 generation as shown in Table 1, the ratio between sterile plants with $Ta1$ gene and partial fertile plants without $Ta1$ gene conforms to 1:1 ratio $(\chi^2 < \chi^2 0.05)$. Whether $Ta1$ gene is located in the D genome or not, the segregation ratio of fertility in F_1 should be 1:1. Because the Taigu male sterile wheat (AABBDD) cannot set seeds by selfing, $Ta1$ gene constantly

Table 1. The fertility of F_1, BC_1, BC_2, and BC_3 of Taigu male sterile wheat with Durum wheat

generations	sterile plants with Tal gene	fertile or partial fertile plants without Tal gene	ratio of sterile/ plants fertile	the place of experiments
F_1	61	56	1:0.9	green house in Dongsheng field in Qinghai Academy of Agricultural Sciences
BC_1	54	144	1:2.7	green house in authors' institute
BC_2	98	356	1:3.6	green house in authors' institute
BC_3	17	166	1:9.8	net chamber in authors' institute

Table 2. Relation of BC_3 fertility to D genome constitution

fertility	No. of plants studied	chromosome				
		14″	14″+1′	14″+2′	14″+3′	14″+4′
fertile	109	86	11	7	3	2
sterile	15	0	9	2	2	2

exists in the heterozygous state, so it always produces two kinds of female gemetes, i.e., one half with $Ta1$ gene, another half without $Ta1$ gene. In F_1 hybrids of Taigu wheat pollinated by tetraploid durum wheat, certainly there should be nearly half sterile plants carrying $Ta1$ gene and half partial fertile plants carrying no $Ta1$ gene too; this has no relation to D genome. The F_1 generation was all pentaploids and their gametes are mostly aneuploids with very low or no viability, so F_1 plants without $Ta1$ gene were partially but not completely fertile.

In Table 1 the fertility segregation of backcrosses did not conform to 1:1 ratio. In BC_1, the ratio between sterile plants and fertile (partial fertile) plants was 1:2.7. While those of BC_2 and BC_3 were 1:3.6 and 1:9.8, respectively. This segregation ratio is attributed to the ratio between two kinds of female gametes:$Ta1$-gene-carrying gametes and gametes without $Ta1$ gene produced by the sterile Taigu wheat. In F_1 sterile plants, the chromosomes of A and B genome formed 14 bivalents but the chromosomes of D genome became univalents. If $Ta1$ gene is on a chromosome of D genome, the gene may be missing as the $Ta1$-gene-carrying univalents disappear, so that BC_1 generation would produce much more fertile plants than the sterile plants. If $Ta1$ gene is not the chromosome of A or B genome, segregation for fertility will always keep 1:1 ratio. The same situation would happen in BC_2 and BC_3 but the chance of univalent loss increases with the advances of backcrossing generation. So we think the $Ta1$ gene must be on a chromosome in D genome.

2. Cytological study of backcrossing progenies

We made cytological studies on backcross BC_1, BC_2, and mainly on BC_3 progenies (Table 2) while investigating the fertility of F_1 hybrids and backcrossing progenies. The data in Table 2 show that all sterile plants had univalents from D genome and those which had exclusively 14 bivalents of A and B genomes and no univalents of D genome were fertile. We crossed $14'' + 1'$ sterile BC_2 plants (plate IV, 3) with durum wheat and the results proved that all 5 sterile plants had the $14'' + 1'$ chromosome constitution while all 43 fertile plants had only $14''$. Consequently, if the univalent of D genome existed, the plants would be sterile; on the contrary, once the univalent was lost, the plants became fertile, indicating $Ta1$ gene is located on one chromosome of D genome.

$Ta1$ gene location on chromosome
1. Ditelosomic test

From the results of genome study and the $14'' + 1'$ sterile plants, the telocentric test can be used for further studies on $Ta1$ gene location[9]. Because $Ta\ 1$ gene is located on the unique univalent in $14''+1'$ sterile plants, ditelosomic lines of D genome (at least one arm in ditelosimic line of each chromosome is available) can be used to cross with $14''+1'$ sterile plants, then $Ta1$-gene-carrying chromosome can be defined by examing F_1 fertility and chromosome constitution. If chromosome constitution of F_1 sterile plants is $14''+1t''+6'$, $Ta1$ gene will situate on the chromosome corresponding with the ditelosome employed. If chromosome constitution of F_1 sterile plants is $15''+5'+t'$, $Ta1$ gene will not be located on the chromosome of the telosome concerned.

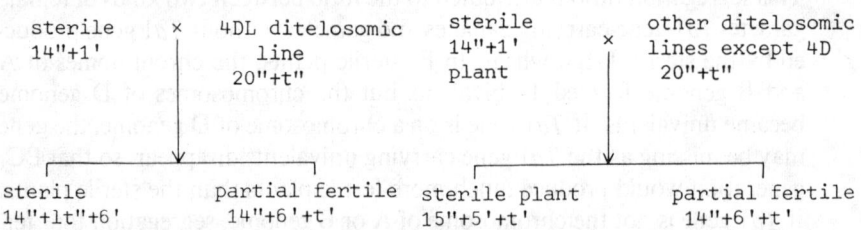

Fig. 4 Diagram of telosomic test   Fig. 5 Diagram of telosomic test

When $14''+1'$ sterile plant was crossed with ditelosomic lines: 1DL, 2DS, 3DL, 5DL, 6DL, 7DL, and 7DS, the 15 bivalents, 5 univalents and 1 monotelosome were found in F_1 sterile plants at MI as Fig. 5, which showed $Ta1$-gene-carrying univalent of D genome had not paired with above telosomic, we were sure that $Ta1$ gene was not on the chromosomes of these ditelosomes. When $14''+1'$ sterile plant was crossed with 4DL ditelocentric line, sterile F_1 had 14 bivalents, one heteromorphic bivalent, and 6 univalents at MI (Fig. 4, Plate IV, 4) $Ta1$-gene-carrying univalent of D genome paired with 4DL telosome to form a heteromorphic bivalent, i.e. $Ta1$ gene is located on 4D chromosome.

2. Observation on the shape, size and arm ratio of the $Ta1$-gene carrying chromosome

Since $Ta1$ gene is located on the univalent of the $14''+1'$ sterile plant, the shape, size and arm ratio of $Ta1$-gene-carrying chromosome can be distinguished at meiosis in the $14''+1'$ sterile plant. The results of the observation demonstrated that the $Ta1$-gene-carrying univalent and the univalent of 4D monosomic line are similar in shape and size. The arm ratio of $Ta1$-gene-carrying chromosome was measured 1.83 (Plate IV, 5) at meiosis while Sears tested the ratio of chromosme 4D was 1.8[8]. Again the observation was well conformed to the result of telocentric test.

Telocentric analysis of $Ta1$ gene

We made the cytological study of 12 telosomic lines of 7 chromosomes in D genome, in which both arm telocentric lines of chromosome 2D, 3D, 4D, 6D and 7D were obtained. The results (in Table 3) demonstrated that in short arm or long arm telocentric family lines of chromosome 2D, 3D, 6D and 7D, the fertile plants had both heteromorphic disomic and normal disomic, and agreed with 1:1 ratio. This case occured only in the non-critical family lines, in otherwords it indicated $Ta1$ gene is not situated on the chromosome 2D, 3D, 6D and 7D. Chromosome 4D is the exception. In the long arm and short arm telocentric lines of chromosome 4D, the ratio of heteromorphic disomic and normal disomic in the fertile plants was significantly deviated from the 1:1 ratio. This is the case in the critical family line, indicating $Ta1$ gene is located on the chromosome 4D. For chromosome 1D and 5D, We checked only long arm telocentric line but they had 1:1 ratio of normal and heteromorphic disomics, i.e. $Ta1$ gene would scarcely exist on these two chromosomes.

129 fertile plants in 4DL telocentric family line were observed, among which, except for one monosomic, 128 plants were the disomics with heteromorphic bivalent. Their sterile plants were checked also; 81 plants were all normal disomics in chromosome constitution. This means all fertile plants were heteromorphic disomics while all sterile plants were the normal disomics (except one). This is the case which only occured in the chromosome arm carrying no $Ta1$ gene. As to the monosomic fertile plants, judging from the chromosome shape and size, it might be due to the fertilization with a faulty gamete which missed long arm of 4D telo at meiosis.

Among the 4DS family line, 182 fertile plants were observed, 56 were disomic plants, 118 were heteromorphic disomic, 8 were monosomic. Besides, 51 of 76 sterile plants examined were normal disomics, 22 heteromorphic disomics and 3 monosomics. The ratios of normal and heteromorphic disomics both in fertile and in sterile plants did not agree with 1:1 ratio. This happened only in the critical telocentric family line (Fig. 2). The normal disomics appeared in the fertile plants and the heteromorphic disomics in the sterile plants would be the recombinants resulted from the crossingover

Table 3. Chromosome constitution of fertile plants of (Taigu wheat × Ditelo.) F_1 sterile × normal disomics

Ditelo plant number chromosome constitution	$20'' + 1t''$1)	$21''$	$20'' + 1'$	χ^2 (1:1) ($\chi^2_{0.05} = 3.84$)
1DL	21	21	1	0.02
2DL	11	15	1	0.35
2DS	23	26	1	0.08
3DL	23	20	5	0.09
3DS	13	15	1	0.04
4DL	128	0	1	126.00
4DS	118	56	8	21.39
5DL	47	44	4	0.04
6DL	22	19	2	0.10
6DS	25	30	7	0.29
7DL	12	15	0	0.15
7DS	23	25	3	0.02

1) annotation: $20'' + 1t''$ includes $19'' + 1t'' + 1'$ et al.

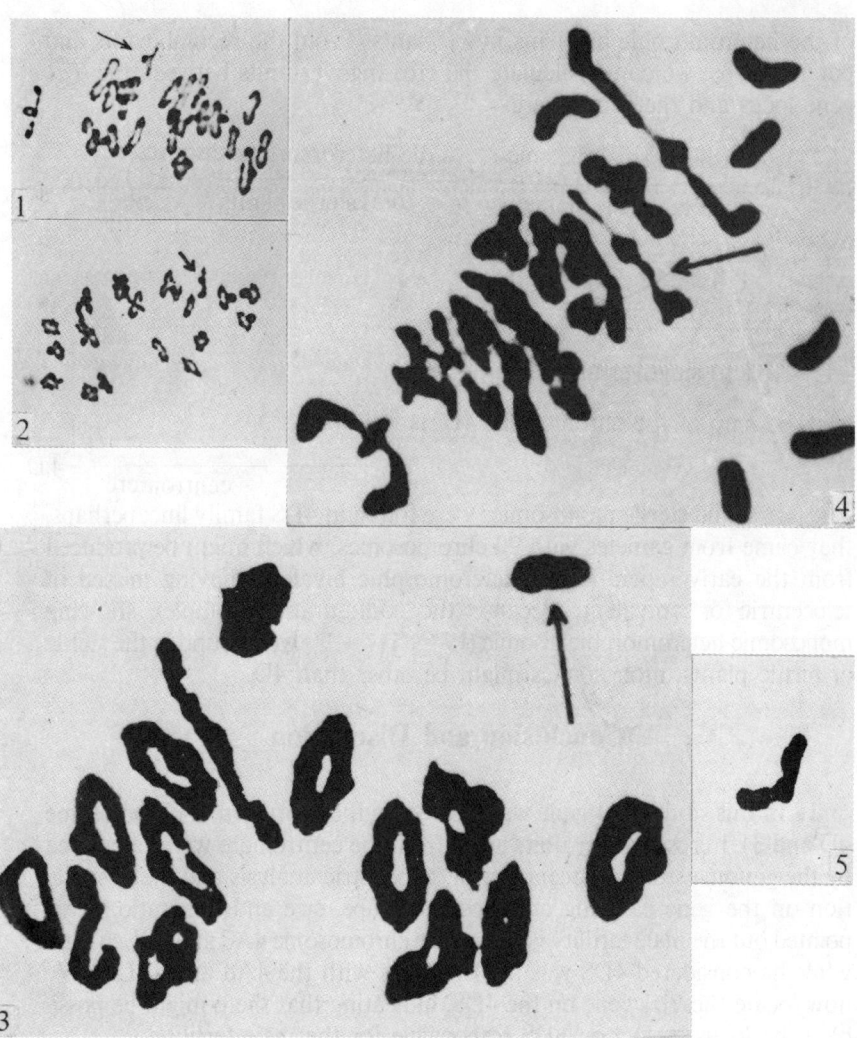

Plate IV

1. Ditelosomics (20" + t").
2. Heteromorphic disomics (20" + 1t").
3. Sterile plants: 14" (AABB) + 1'(D).
4. Sterile plants: 14" + 1t" + 6'.
5. Univantent-4D at AII.

of the heteromorphic bivalents in F_1 plants. From the recombinants and parental type, we could calculate the crossingover units between the $Ta1$ gene locus and the centromere.

$$S = 1/2 \left(\frac{\text{fertile normal disomics}}{\text{total fertile plants}} + \frac{\text{sterile heteromorphic disomics}}{\text{total sterile plants}} \right) \times 100$$

$$= 1/2 \left(\frac{56}{174} + \frac{22}{73} \right) \times 100$$

$$= \frac{32.1 + 30.1}{2} \times 100$$

$$= 31.1 \text{ (crossingover units)}$$

The map of the chromosome 4D as follows:

$$\underset{\text{centromere}}{\bullet} \overset{Ta1}{\underset{}{\rule{3cm}{0.4pt}}} 4\text{D}$$

The fertile and sterile monosomics were found in 4DS family line; perhaps, they came from gametes with 20 chromosomes, which might be produced from the early seperation of heteromorphic bivalents having missed of telocentric or univalent. Because the configuration complex showing monosomic-heteromorphic disomic ($19'' + 1t'' + 1'$) were found in the sterile or fertile plants, monosomics might be other than 4D.

Conclusion and Discussion

1. In this study, $Ta1$ gene was located on the short arm of chromosome 4D and 31.1 crossingover units apart from the centromere were calculated by the genome study, telocentric test, telocentric analysis and the observation on the gene carrying chromosome shape, size and arm ratio. Sears pointed out the male fertility genes on the chromosome 4Ad and 4BL. Meanwhile he considered 4DS was homeologous with the 4Ad and 4BL[4]. We now locate the $Ta1$ gene on the 4DS, indicating that there might be possible gene locus or loci on 4DS responsible for the male fertility.

2. In the progenies of backcross of $14'' + 1'$ sterile plants with durum wheat, when ever the univalents from D genome were present, the plants were still sterile; once univalents were absent, they became fertile. The fertile plants ($14''$, AABB) never carry the allell $ta1$ corresponding to $Ta1$ gene, i.e. the existence of $Ta1$ gene could suppress all gene functions controlling the male fertility.

3. The genome study coupled with the telocentric analysis were applied

in this experiment; this is a new approach for gene location in wheat, which will not only reduce a lot of work but also may reaffirm the results in two ways and raise the accuracy of the test as well. The 14″ + 1′ sterile plants from the genomic study may be directly used not only to observe the shape, size and arm ratio of chromosome carrying $Ta1$ gene, but also to provide a rather simple approach for telocentric mapping, in which 14″ + 1′ sterile plants were crossed with each ditelocentric lines of D genome. The critical chromosome carrying $Ta1$ gene can be confirmed by checking the chromosome composition and fertility of their progenies.

4. Since $Ta1$ gene is on the D genome, the gene can be transferred into wheat relatives or new species containing D genome in order to make full use of these wild species or improve the new species.

5. $Ta1$ gene is on the chromosome 4D, while blue aleurone gene is on the 4E in *Agropyron*, 4D and 4E are somewhat homeologous, which opened way to link blue gene and $Ta1$ gene together by inducing homoelogous chromosome pairing, If both genes are linked up into one chromosome, sterility and fertility can be distinguished by seed colour[12]. It would be of great importance in heterosis utilization, recurrent selection, and theoretical reseach.

References

[1] Dang, K. D. et al., 1982, Discovery and Determination of A Dominant Male-sterile Gene and Its Importance in Genetics and Wheat Breeding. *Scientia Sinica* (Series B). **25**(5): 508-516.

[2] 刘秉华,1982,小麦遗传研究和育种中应用非整倍体的几个问题,农学文摘,（1）:1-7。

[3] Sears, E. R., 1953, Nullisomic analysis in common wheat. *Amer. Nat.* **87**: 245-252.

[4] Sears, E. R. and Lotti M. S. Sears, 1978, The telocentric chromosomes of common wheat. Proc. 5th Int. Wheat Genet. Symp. 389-407.

[5] Driscoll, C. J., 1974, IAEA, Polyploidy and Induced Mutations in Plant Breeding, 139-142.

[6] Driscoll, C. J., 1978, Induction and Use of the "Cornerstone" male-sterity in wheat. Proc. 5th Int. Wheat Genet. Symp. 499-502.

[7] Maan, S. S., 1966, Development and Use of an aneuploid set of male sterile Chinese Spring Wheat of *Triticum timopheevi* Zhuk Cytoplasm. *Canadian Journal of Genetics and Cytoplogy.* **8**: 398-403.

[8] Sears, E. R., 1954, The aneuploids of common wheat. *Missouri Agr. Exp. Sta. Res. Bull.* (572): 1-59.

[9] Sears, E. R., 1966, Chromosome mapping with the aid of telocentrics. Proc. 2nd Int. Wheat Gent. Symp. 2: 370-380.

[10] Kimber, G. and Sears, E. R., 1980: Uses of wheat aneuploids. In Polyploids: Biological Relevance. Plenum Publishing Corp. New York.

[11] Roba, T. and K. Tsunewaki, 1978, Mapping of the s and ch2 genes on chromosome 3D of common wheat. Wheat Information Service. (45-46): 18-20.

[12] Sears, E. R., 1973, Agropyron-wheat transfers induced by homoeolygous pairing. Proc. 4th. Int. Wheat Gent. Sypm. Columbia, Mossouri, pp. 191-199.

THE CYTOLOGICAL AND HISTOLOGICAL INVESTIGATION OF THE ABORTIVE PROCESS

Li Xiangyi Deng Jingyang

(*Institute of Crop Breeding and Cultivation, Chinese Academy of Agricultural Sciences*)

This is a continuation of the cytological studies on the Taigu male sterile wheat by Deng Jingyang et al. The objective of this study is to look into the mechanism of the male sterility of this material in order to make better use of the *Ta*1 gene.

Materials and Methods

The materials used were the sterile and fertile plants segregated from the open-pollinated Taigu wheat, as well as a check cv. Nongda 139. Seeds from the sterile plants were sown into 50 earlines on 25th of September, 1980, and inter-planted with Nongda 139.

Sampling materials were taken from late April of the following year, when the sterile plants and fertile plants could be identified by checking the size, appearance and colour of anthers in the young spikes of main culm at booting stage. At this time the young spikes of robust tillers of each plant were collected for the cytological study. Samples were taken every other day until flowering and fixed in carney's solution and squashed in acetocarmine. For paraffin sectioning, FAA fixation and Haematoxylin stain were used. The typical section slides were photographed.

The investigations were repeated with materials grown in the greenhouse in winter of 1980 to spring of 1982. The results were similar with those from the materials grown in the fields.

Results and Analysis

The development of pollen and anthers of fertile plants
1. Pollen development

The mature pollen mother cells are arranged along the inner layer of tapetum and enclosed by calli. PMCs produce tetrads at meiosis. The microspores turn round after being released from the callus and develop into monokaryotic pollen enclosed by bilayered cell wall with clear germ pore. At the late mononuclear stage, pollen undergo a hetero mitosis and form the binucleate cell (a vegetative nucleus and a reproductive nucleus). Then the pollen enter the starch filling stage. Before anthesis, the reproductive nucleus undergoes another mitosis to produce two sickle-shaped sperms, which constitute the trinucleate pollen.

2. Anther development

Wheat anther wall consists of four layers of cells: epidermis, inner wall, middle layer and tapetum. When PMCs start meiosis, the cells of middle layers are squeezed, stretched and gradually decomposed, but the residues still exist until binucleate stage (Plate VII, 1). At early stage, the tapetum cells have large nucleus and thick cytoplasm, having characteristics of typical gland cells. Before meiosis of PMCs, cells in tapetum pass through mitosis to form binucleate cells. At the stage of free spores, tapetum starts to degenerate. At the binucleate pollen stage, the tapetum gradually disappears. At flowering, only inner wall and epidermal cells are left in the anther wall. The inner wall cells between the anther compartments have secondary fibrous growth and anthers dehisce logitudinally.

The male abortion

The male abortion in the sterile plant mainly occurs in the formation of microspores or after the release of microspores, which were decomposed rapidly and completely, so that the anther is called nonpollen type

1. The rapid decomposition of microspores after their being released

Observations revealed that there were various types of microspores in one anther. In most anthers, pollen mother cells could pass through meiosis and give rise to tetrads (Plate V, 14). The micropores released from tetrads could not increase much in size. Some appeared round but had not formed wall and germ pore and rapidly degenerated (Plate VII, 5). In other cases two young microspores stuck together in a tetrad and looked binucleate in the microscope (Plate V, 12). We also observed that many dyads did not divide any more and became very large (Plate V, 9-10). In some anthers, dyads were counted as high as 50-70%. The dyads sometimes had karyokinesis to become binucleate. There also existed aberrant tetrads, triads. The microspores were not uniforminsize and were therefore called aberrant microspores (Plate V, 11-13), and they died just after release. The calli of some tetrads could not be dissolved and young microspores were gradually decomposed in the mother calli.

2. The abnormal differentiation and early development of pollen mother cells

The male abortion is not a instantaneous degradation process which occurs suddenly after the release of microspores. We discovered that the mitosis prior to meiosis was not complete or even did not take place. For instance, in the cross section of the anthers in sterile plants, only 4-6 pollen mother cells were found, while in the fertile anthers, there were 6-8 PMCs. In the longitudinal section of anthers, there were only one line of PMCs containing 22-24 cells per line in some sterile anthers, while there were 2 lines, about 40 PMCs per line in fertile anthers. Another instance is that in the cross section of some sterile anthers, we found some cells directly developed into PMCs without the radial split at mitosis, moreover, daughter cells did not separate from each other and entered meiosis prematurely. (Plate V, 2)

3. The male abortion can be seen at meiosis

Around the begining of meiosis, the tapetum in a few anthers had aberration and degradation, meantime the decomposition of PMCs was observed. Plate V, 3 illustrates that the multi-nuclear plasmodia of the tapetum underwent anomalous growth and PMCs were squeezed to break and die. In some anther cells, PMCs were fused into a mass and dead (Plate V, 4). In some anthers, shortly after the beginning of meiosis when the tissue of tapetum was almost decomposed, PMCs stopped dividing and swiftly disintegreted (Plate V, 5). Besides, in some anthers or anther compartments PMCs only passed mitosis and the resulting daughter cells were also quickly disintegreted (Plate V, 6).

4. Extremely aberrant meiosis

In most of anthers, the meiosis from leptotene to zygotene were generally normal. However, there was rather severe adhesence of chromosomes, 40% of which were arranged irregularly along the equatorial plate. Some chromosomes were far away from the equatorial plate (Plate V, 7). Several laggards were observed at AI(Plate V, 8). Besides, in dyads the spindle fibres still remained. At AII, some dyads often divided asynochronously resulting in the anomalous quartets and triads (Plate V, 11). There also existed mitotic division (Plate V, 6). Small number of PMCs had not undergone cell division and continued to remain and were disintegrated together with microspores (Plate VII, 7).

5. The pollen abortion occurred relatively late in some anthers

Since late collapse of tapetum happened in a few anthers, microspores could develop into uninucleale pollen, even binucleate pollen. The volume of which had increased several fold. Bilayered cell wall and germ pore were formed. These crowded the compartment with limited volume. Then they did not grow anymore. The nuclear and cytoplasm were almost decomposed at the same time. Some uninucleate pollen in the process of decomposition were considered as nonnucleus pollen. At last, only a little pollen wall remained in the anther compartment (Plate VI, 3-7).

6. Uneven microsporogenesis and development of microspores in anthers and anther compartments

We found that some anthers had 3 empty chambers but PMCs still grew in the other chamber. Plate VI, 1 illustrates four anther chambers: one is empty, one is full of cells fused together to form multinucleate plasmodia. In the other two chambers, microspores were rapidly degenerating after being released from meiosis. The majority of microspores were uninucleate and some binucleate. In a squash slide of one floret, we observed some PMCs were at meiosis while some microspores were already degenerated after meiosis (Plate V, 8). In a cross section of one floret, 11 anther compartments were empty but the remaining one was full of the uninucleate pollen (Plate VI, 3). Plate VII, 6 demostrates that microspores were degenerated in two chambers of one anther, but uninucleate pollen still lived in the other two. These indicate that several abortive phenomena occurred in different ratio in one floret or even in one anther.

The relation between the early anther tapetum decomposition and the microspore degeneration in the sterile plants

The aberration or decomposition of tapetal cells of anthers in the sterile plants always occurs before the degeneration of the microspore or PMCs. Cytological evidences are shown in Plate V, 3-5. In many anthers the protoplasm of tapetum became vacuolated quickly and the whole tissue completely collapsed before the microspore released in harmony with the critical stage of microspore degeneration (Plate VII, 2-5). Only when the tapetal layer develops normally or degeneration takes place late, could the microspores develope into uninucleate pollen (Plate VI, 2-4). We discovered the tapetal cells started to become sharply vacuolated early at prophase I of meiosis (Plate VII, 2). At the dyad stage, the tapetal cell was highly vacuolated with very little cytoplasm and the nucleus was squeezed aside by a big vacuole (Plate VII, 3). Before the release of microspores, the nucleus and cytoplasm of the tapetal cell disappeared, the remaining fiberous cell wall was faintly visible, hence the microspores were aborted just after being released or a little later (Plate VII, 4–5). Plate VII, 6 shows that in two of 4 chambers in an anther, tapetal cells have developed normally, so microspores developed into the monokaryotic pollen; in the other two, tapetal cells collapsed and development of microspores stopped and rapidly degenerated.

The late degeneration of mid-layer tissue——the prominent feature of anthers in the sterile plants

The middle layer of anthers in the fertile plants generally had degenerated at tetrad stage, only some remaining nod-shaped nuclei were adhered to the outer layer of tapetum. Some cytoplasm remains still stuck to inner layer even until the formation of binucleate pollen (Plate VII, 1). The mid-layer of anthers in the sterile plants, on the contrary, kept intact after the decomposition of tapetal layer or during the microspore degeneration (Plate VII, 7). Only a very few microspores became vacuolated. However, we found in one or two anther chambers that the mid-layer was very thick and the tapetal layer was pressed into the chamber so that a large number of PMCs were damaged. Plate VII, 7 shows that the delayed degeneration of the mid-layer could effect the development of tapetal layer and PMCs. We made microchemical analysis and the results show the contents of the important compounds like ribonucleic acid, protein (by Ninhydrin-schiff staining), polysaccharide (by PAS staining) and fat (by sudan III staining) in mid-layer of anthers of sterile plants were remarkably higher than those in fertile plants. The above-mentioned staining reactions in other tissues (PMCs, tapetal layer, inner layer,) were much weaker than those in fertile plants, indicating the anther mid-layer in sterile plants stored or utilized the limited nutrients from the vascular bundle which helped to accelerate the deterioration of tapetum and mirospores. The results obtained from the microchemical and cytoloical studies were the same. In addition, the inner-layer also become vacuolated early (Plate V, 9; Plate VII, 7). The thickened secondary fibrous layer was observed in late stage. Anthers did not dehisce at anthesis. After the chamber became empty, inner wall and epidermal cells became very large and extended inward so that chambers were compressed to be very small.

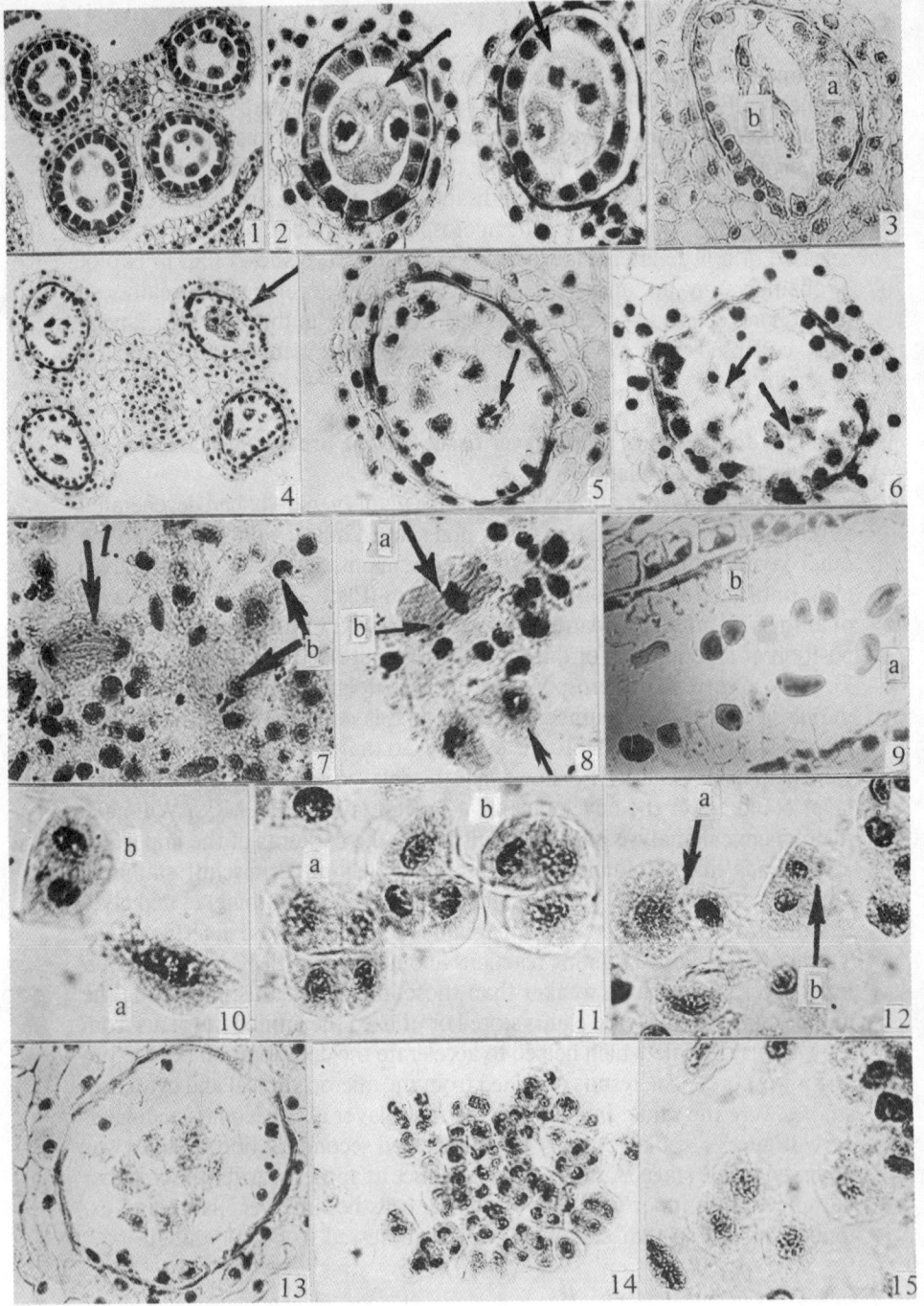

Plate V

1. (× 67): Prophase I of meiosis in the fertile plants. 2–12 Referring to the fertile anthers.
2. (× 160): Some cells directly develop into PMCs without the longitudinal mitotic division and start meiotic division at the callus stage.
3. (× 160): (a) The aberrant growth of the protoplasm mass in the tapetal layer; (b) the PMCs have been squeezed to death.
4. (× 67): PMCs are fused and dead in some anther compartment prior to meiosis.
5. (× 160): At prophase I, the tapetal tissue has been basically degenerated and PMCs stop dividing and rapidly decomposed.
6. (× 160): PMCs rapidly decomposed after mitosis.
7. (× 330): Metaphase I, (a) the chromosomal adhensive phenominum; (b) the irregular arrangement of chromosomes at the eqatorial plate.
8. (× 330): (a) Some laggards at anaphase I; (b) Dyads after release from the callus and PMCs at telophase I.
9. (× 160): (a) Very big dyads; (b) The thorough degeneration of tapetal layer.
10. (× 550): (a) The degeneration of dyads after release; (b) Two young microspores are linked together like "binucleur spores".
11. (× 500): (a) The aberrent tetrads; (b) Triads.
12. (× 330): (a) The microspores with irregular shape and different size; (b) Two young non-split microspores.
13. (× 160): The abnormal microspores.
14. (× 160): Some tetrads in a floret.
15. (× 330): Microspores released from tetrads are degenerating rapidly.

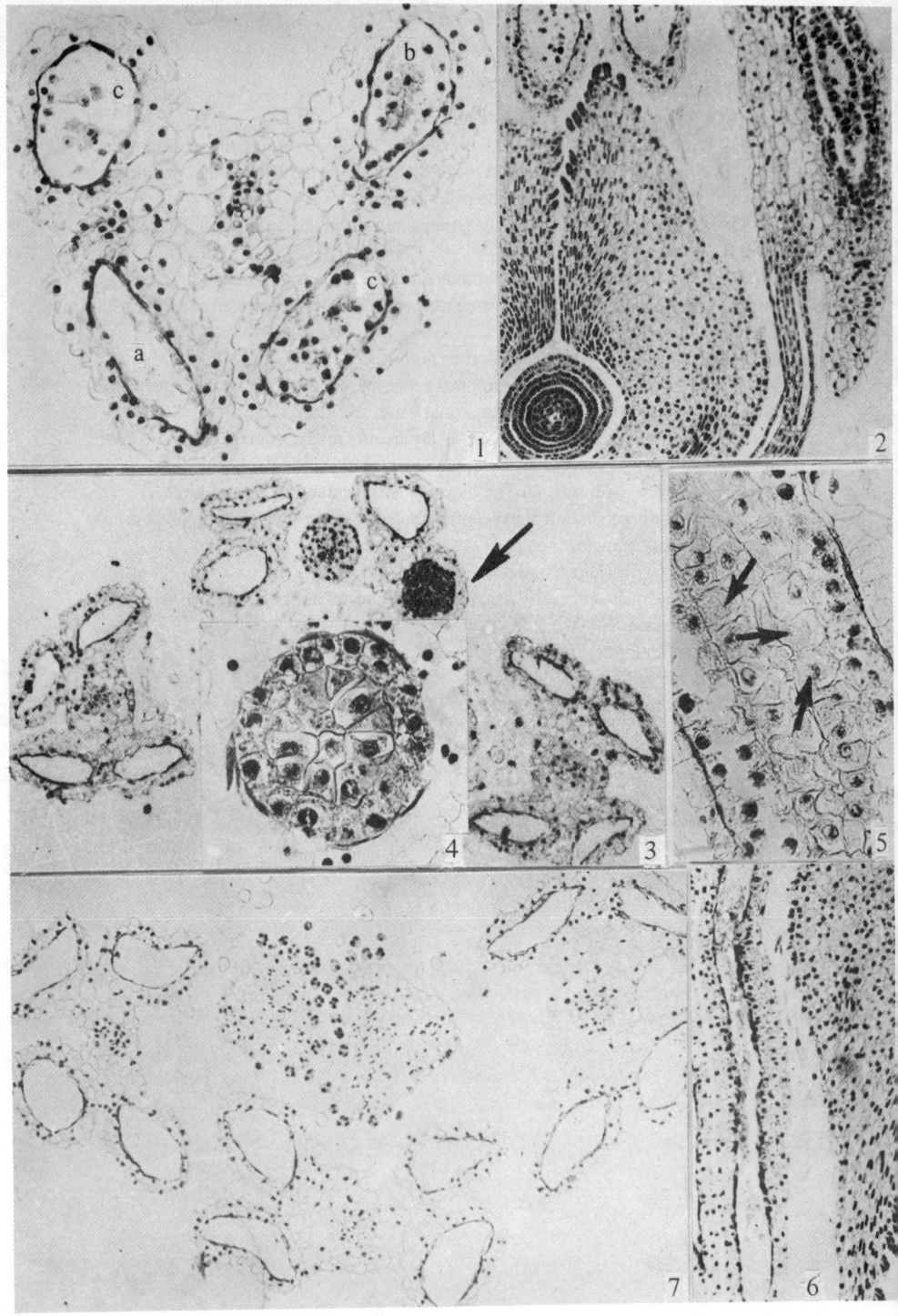

Plate VI

1 - 7 The anthers of sterile plants.
1. (× 67): In one anther, (a) one empty compartment, (b) another compartment with multinucleonate, (c) uninucleate microspores in the rest two compartments.
2. (× 25): Microspores among anthers of a floret are at different stages.
3. (× 25.6): There are 11 empty anther compartments and still normal pollen in one compartment.
4. (× 160): The enlarged picture of one compartment from 3, the uninucleate pollen is full of the rather limited volume.
5. (×200): The cytoplasm and nucleus are decomposed simultaneously. Some uninucleate pollen in the process of decomposition are often considered as "nucleusless pollen".
6. (× 33): No pollen in all anther compartments of a floret at heading.
7. (× 25): A vertical section of a floret whose microspores have broken down.

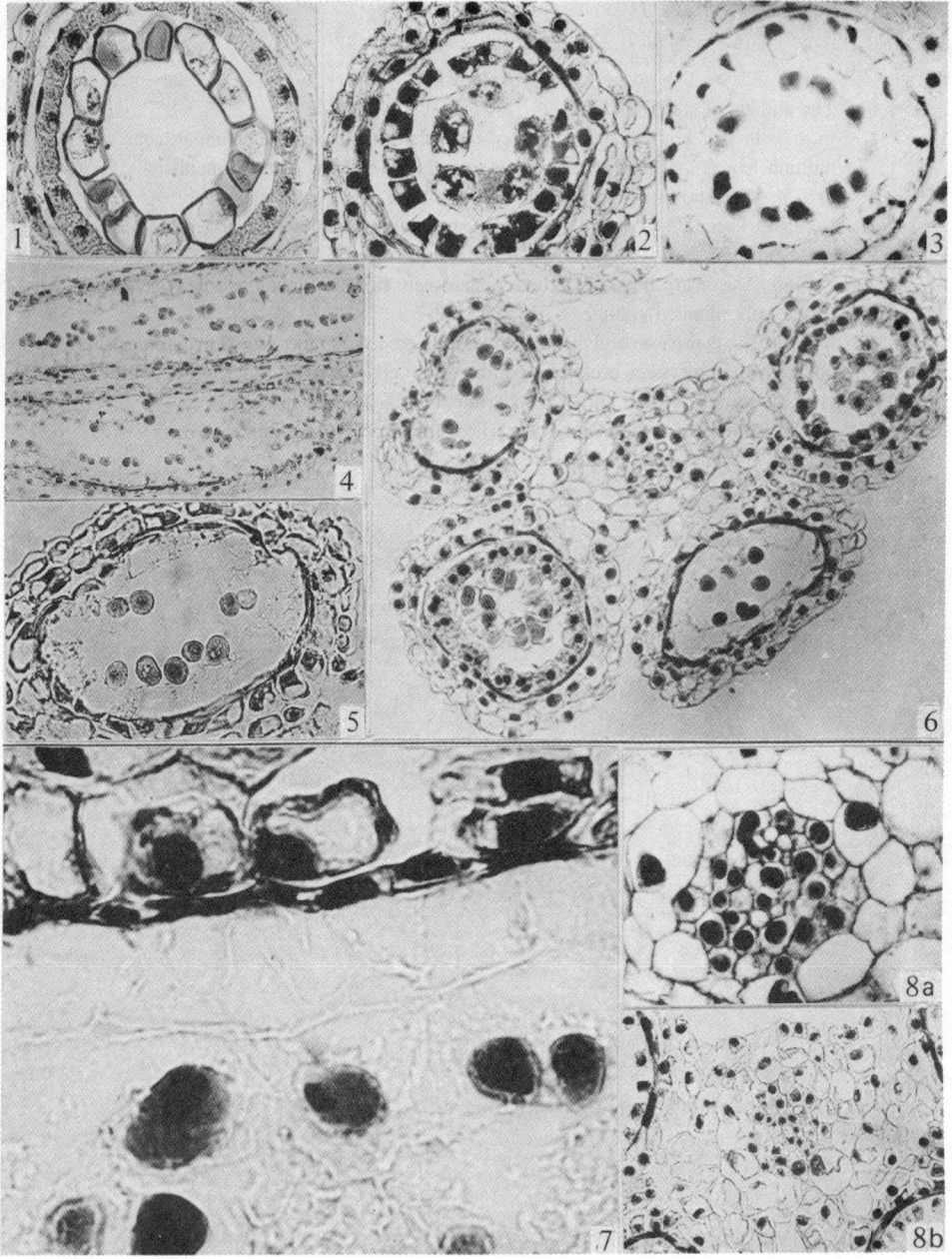

Plate VII

1. (× 160): At the binucleate pollen formation stage in the fertile plants, the tapetum seems decomposed, the remains of mid-layer is still visible.
2-4.The process of the tapetum protoplasm degeneration in the sterile plants.
2. (× 160): Prophase I.
3. (× 160): Dyad stage.
4. (× 67): Microspore before release.
5. (× 160): Completely decomposed tapetum and aborted microspores after release.
6. (× 67): In one anther, two anther compartments are normally developed with the uninucleate tic pollens; in other two anther compartments, tapetum has been decomposed and microspores stop developing and to degenerate.
7. (× 268): In the process of microspore abortion after degeneration of tapetum; the whole tissue of mid-layer still remains, highly vasculated inner layer cells.
8. a. (× 160): Fertile anther vascular bundle; b. (× 67): irregular arrangement of vascular bundle sheath and Ill-developed cells in the sterile anthers.

Abbreviation used in Plate VIII-XI.
Cu: cuticle; Ep: epidermis; En: endothecium; Ta: tapetum; M-L: middle-layer; Nu: nucleus; ER: endoplasmic reticulum P: plastid; M: mitochondrium; V: vacuole; N-M: nuclear membrane; C-M: concentric circle-shaped membrane structure; O-P: osmiophilic particle; A-V: autophagic vacuole; Ex: exine; Sp: spherosome; U: Ubisch body; C-S: compound structure.

The aberrant morphology of vascular tissues in sterile plants

Clear and visible sieve tubes and vessels were observed in vascular bundles of the fertile plant anthers. Their living cells had large nuclei and thick cytoplasm. The parenchyma cells of the vascular bundle sheath were arranged regularly around the vascular strand (Plate VII, 8a). Although there were vascular strands and vessels in the sterile plant anthers, the texture of the vascular bundles was irregular, their living cell cytoplasm was condensed, cell wall and protoplasm were separated and the nucleus was small and looked disintegrated. The parenchyma cells were arranged in disorder with some cell cytoplasm condensed and deposited aside. Some parenchyma cells were remarkably swollen and cytoplasm was anomalously growing (Plate VII, 8b). Apparently certain physiological obstacle in the anther vascular tissue retarded the transportation or exchanges of substances. We also discovered the early aberration of living cells in anther vascular tissue was closely related to the degeneration of PMCs and deformation of anther wall tissue like tapetum. Because physiological activities of these live cell were disordered, the conducting function of vascular strands was upset so as to speed up the male abortion and the forced death of anthers.

Discussion and Conclusion

The male abortion of Taigu male sterile wheat controlled by the *Ta*1 gene possesses the following features: firstly, the large number of abortion phenomena which take place in the process of microspore formation and after their release; the microspores in only a few anthers or anther sacs, develop to the uninucleate (or binucleate) pollen and then start to degenerate. No matter whether the abortion happens earlier or later, the microspores are destined to decompose rapidly and completely within a very short period. This process is called nonpollen type abortion. Secondly, meiosis is very irregular or does not take place. Thirdly, the microsporogenesis and microspore development among anthers or anther sacs are irregular. Several kinds of abortive phenomena are often found in different proportion in one floret or even in one anther.

The nonpollen type abortion has been reported in the study on the male sterile rice in China[3]. Its features are very similar to those of Taigu male sterile wheat. This research indicates that the abortion of Taigu male sterile wheat is not a result of instantaneous degeneration that happens suddenly after the release of microspores, but is an inevitable consequence of extremely aberrant formation process of microspores. Starting from the abnormal differentiation of pollen mother cells, there appears lack of the callus protection of PMCs in various degrees at interphase, then meiotic asynchronism and abnormality in over half of the dyads and aberrant microspores, the large number of abnormal tetrads and finally, many microspores are decomposed just after their release. Among them there are also some tetrads which can pass through a very short period of development and then abort. The observation of the dynamic processes and anther wall indicates that the expression of *Ta*1 gene starts very early although the collapse of microspores may occur one after another. And male sterility is always related to the nutrition supply system.

The tapetum tissue is generally regarded as the "nurture tissue" in the development of anthers[5, 6]. Data on genic sterility or cytoplasmic male sterility in different plants indicate that, while male sterility may take place from the PMC differentiatin till the release of tetrads, the tapetum abnormality is always earlier than the microspore abortion[5, 7, 9]. However there are some contrary opinions[10]. From Plate VII,6 we can see that, the tapetum in two of four anther sacs in one anther develops almost normally and the microspore can develop into the uninucleate stage. The tapetum in other two anther sacs has collapsed and the microspores stop their development and then decompose rapidly. This proves that the untimely decomposition of the tapetum plays the decisive role in the degeneration of micropores.

The mid-layer tissue of the anthers in the normal fertile plants has generally decomposed at the tetrad stage whereas, in Taigu male-sterile wheat, a complete anther mid-layer tissue can be generally maintained after the collapse of tapetum and during the degeneration of microspores. We also have observed the mid-layers are very thick and vacuolated in a few of the anther sacs. The result of microchemical analysis indicates that the contents of the important compounds such as ribonucleic acids, proteins, polysacchasides, fats, etc., in the mid-layer tissue of the sterile plants are much higher than those of ferile plants. We suppose that the abnormal mid-layer tissue may store or utilize the limited nutrients from the vascular bundles of anther-comipartments instead of transporting nutrients to the tapetum, thus accelerating the decomposition of tapetum. This phenomenon has not yet been reported in literature. On the other hand, the mid-layer in the normal anther acts as a nutritional tissue, that is, it maintains for a time the connection between the anther-compartment and the tapetum; then the degradated substances are absorbed by the tapetum.

There have been some reports about the abnormality of the anther-compartment vascular bundle in the male sterile plants[4,8,9]. This paper demostrates that the protoplasm of many parenchyma cells in vascular bundle, sheath and anther-compartment of the anther in the Taigu male sterile plants coacervates and deteriorates. This physiological breakdown in the living cells destroys or severely damages the transporting function of the anther vascular bundle. It has an especially remarkable effect on the transportation of assimilates in sieve tubes and destroys or weakens the transport of plasmodesmata in a short time., Deng et al. have indicated that the stagnation of anther growth and the male abortion occur almost synchronously[1]. Our experiment has proved that the stunted vascular bundle tissue of the anther and the early aberration of its living-cells play a decisive role in the acceleration of the male abortion and the premature death of the anther. We infer that the dominant male sterile gene $Ta1$ can express itself not only in the sex cells, but also in the somatic cells of the stamen.

References

[1] 邓景扬等，1980，小麦显性雄性不育基因的发现与利用—太谷不育小麦鉴定总结，作物学报，6（2）：85-98。

[2] 北京大学生物系植物遗传育种专业，1976，雄性不育与雄性能育小麦花药和花粉发育的细胞形态学观察，植物学报，18（2）：141-149。

[3] 湖南师范学院生物系，1978，水稻雄性不育的花粉败育途径，中国农业科学，(3)：1-7。

[4] 中山大学生物系遗传组等，1976，作物三系生物学特征的研究Ⅲ，几种水稻雄性不育类型的花粉形成与发育的细胞形态、代谢障碍和药隔维管发育的比较研究，遗传学报，3（2）：119-127。

[5] Laser, K. D. and N.R. Lersten, 1972, Anatomy and cytology of microsporogenesis in Cytoplasmic male-sterile angisperms. *Bot. Rev.*, 38: 425-454.

[6] Gotteghalk, W. and M. L. H. Kaul, 1974, The genetic control of microsporogenesis in higher Plants. *The nucleus*, 17(3): 133-166.

[7] Bowman, D. t. et al. 1978, Analysis of a dominant male-sterile character in upland cotton. I. Cytological studies, *Crop Science*, 18(5): 730.

[8] Joppa, L.R. 1966, Pollen and anther development in cytoplasmic male sterile wheat (*T. aestivum* L.). *Crop Science*, 6: 296-297.

[9] Chauhan, S. V. S. and Toshiro Kinoshita, 1980, Anther ontogeny in genic, cytoplasmic and chemically induced male sterile plants. *Japan. J. Breed.*, 30(2): 117-124.

[10] Jain, a. K, A. K., S.Rathore, S. V. S. Chauhan and Tashiro Kinoshita, 1981, Anther ontogeny in an exotic and six indigenous cytoplasmic male sterile lines of wheat (*Triticum aestivum*) possessing *Triticum timopheevi* cytoplasm, *Japan. J. Breed.*, 31(3): 251-260.

SUBMICROSCOPIC OBSERVATIONS ON ANTHER TISSUE AND MICROSPORO-GENESIS PROCESS I

Chen Zhu Xizhao

(Department of Biology, Peking University)

Since 1979, microscopic and electro-microscopic observations have been made on the cytomorphological characters of Taigu genic male-sterile wheat. The author described the critical period and fundamental characters of abortion of anther tissue and pollen in Taigu genic male-sterile wheat. The results could be briefed as follows: 1) In sterile plant, the pollen abortion occurred mainly at the later period of microsporogenesis, i.e. from the formation of tetrad to early microspore stage[1,3] .2) Some irregular phenomena could be seen in the microsporogenesis process; they appeared as an abnormal chromosome behavior in the reduction division process of pollen mother cells in sterile plants[3]. 3) In sterile form, tapetum degeneration took place in the process of meiosis of pollen mother cells[1,3].

On the basis of above work, an attempt was made to study the ultrastructure during pollen abortion. Electro-microscopic investigations were made since spring 1982. This report is a primary conclusion of the recent two years' work.

Materials and Methods

Three materials were observed. They were taken from the Institute of Atomic Energy, Chinese Academy of Agricultural Sciences and the greenhouse, Peking University. All materials were the backcross F_1. The fertile and sterile form of progeny from the same plant were used for comparison.

Light-microscopic examination of aceto-carmine squashes were made for the determination of the developmental stages of anther. Anthers were selected from the first or second flosculus in the spikelet. Anthers at different stages of development from fertile and sterile plant were fixed. 2.5% glutaraldehyde and 1% osmium tetroxide (all were prepared in phosphate buffer pH=7.3) were used as the fixative. The specimens were dehydrated in a graded ethanol series, passed through epoxy propane, then turned into the embedding media—Epon 812 epoxy resin.

Ultrathin sections were prepared with LKB 8800 Type ultrathin microtome, and were stained with lead citrate and uranium acetate, then examined in JEM-6C Type and H-300 Type electron microscopes. At the same time, some thin (1μm) sections were made and stained, then observed directly with light-microscope to determine the development stage of every anther.

Results

Comparative ultrastructural observations on microsporogenesis and its degeneration process in fertile and sterile plants

1. Ultrastructural observation on microsporogenesis process in fertile plant

In fertile form, when a pollen mother cell began its meiosis, it had prominent nucleus and nucleolus, and chromosomes were clearly seen too. Some plastids appeared to be simple in structure or cup-shaped and dumb-bell shaped. Mitochondria appeared a little spherical-shaped and some cristae had developed. Some elaioplasts were characterized by electron-dense contents and large size. Endoplasmic reticulum appeared as short or long tubes in cross section, and sometimes showed in bundle. The pollen mother cell's cytoplasm also was rich in ribosomes and contained a few small vacuoles (Plate VIII, 1). In short, at this time, cell organelles generally remained at the primary phase of development.

When pollen mother cell entered into meiotic division to the stage of metaphase I, and anaphase I, the organelles content showed no significant difference from those of pollen mother cell, but its density in distribution seemed to decrease. At telophase I until dyad, there were some mitochondria, plastids, little vacuoles and endoplasmic reticulum showed clearly in cells (Plate VIII, 2).

The fine structure of tetrad and early microspore had much in common. Many organelles evenly-distributed in the cell, including plastids in various form, spherical-shaped mitocondria, endoplasmic reticulum tubes were short.

In addition, little vacuoles and some similar irregular shaped vacuole structures appeared, and Golgi bodies were rare (Plate VIII, 3). During exine accumulated sporopollenin and formed germinal aperture, in microspore some big vacuoles appeared, longer endoplasmic reticulum tubes distributed near the nucleus, and many developing plastids, mitochondria, elaioplasts and vacuoles were formed and it was also rich in ribosomes (Plate VIII, 4).

2. Ultrastructural observation on microsporogenesis and its degeneration process in sterile plant

In sterile form, some abnormal ultrastructures exhibited in the process of microsporogenesis. When a pollen mother cell entered into reduction division at the stage of prophase I, some concentric circle-shaped endoplasmic reticulum could be seen; the number of circles might be from several to more than ten, and ribosomes arranged densely on the endoplasmic reticulum (Plate IX 1-2). Sometimes, endoplasmic reticulum appeared in bundle. In addition, some osmiophilic particles, with electron-dense contents and irregular forms (Plate IX, 2) and many ribosomes aggregated together. This phenomena were described in our early report[4].

At diakinesis, pollen mother cells exhibited significant vacuolation. At this time, in pollen mother cells there were a lot of big or small vacuoles and irregular osmiophilic particles, sometimes with several undefined spherical structures, which were surrounded by unit-membranes and with different electron transparency and size (Plate IX, 3). The nucleus remained completely or the nuclear membrane was undefined.

After that, the process of microsporogenesis developed in an abnormal way, and presented complex electromicroscopic features. Sometimes, the anther cavity, which was surrounded by a degenerative tapetum layer, was arranged by a circle of well-defined cells, but the greater part of these cells were empty, and in some of them there were several spherical structure with different size and electron-density. Once in a while, fragments of endoplasmic reticulum could be seen (Plate IX, 4). On the other hand, in some anther chambers, no clear boundaries between cells could be seen, only a number of spherical structures, some autophagic vacuoles and the fragments of chromosomes which appeared in part of the anther chamber (Plate IX, 5). There were nothing left in even more degenerated anther cavity, some residues of protoplasts and remains of disorganized organelles could be seen occasionally (Plate IX, 6).

It seems that, in sterile plant, when the pollen mother cell got into diakinesis, the process of microsporogenesis left its normal path and tended toward degeneration with various speed. The time of abortion varied with different pollen mother cells.

Comparative ultrastructure observations on the development and disorgnization of tapetal cells in fertile and sterile plant

1. Ultrastructure of tapetal cell in fertile plant

When a pollen mother cell began its meiosis, anther's wall layers in fertile plant could be seen as in plate X, 1. The wall layers of anther consisted of four layers as follows: epidermis, covered with cuticle; endothecium cells rich with chloroplasts and large vacuoles; flat middle layer, with complete nucleus and large vacuoles; the tapetum layer, next to the middle layer. At this time, tapetum cells with single or double nuclei, dominant nucleolus and endoplasmic reticulum in contact with nuclear membrane could be seen clearly (Plate X, 2). Also visible were endoplasmic reticulum, mitochondria, plastids, a few little vacuoles and many ribosomes distributed in it.

Hereafter, along with the development of tapetal cells, endoplasmic reticulum elongated further and became clear with well defined outline. Meanwhile, the degree of vacuolation increased.

From the formation of tetrad to the release of young microspores, the tapetum development seemed to have attained its full extent (mature stage). At this time, single or double nuclei still remained in the tapetum cell but the content and structure of organelles performed complicated changes and displayed different pictures; in a number of sections we could find tapetal cells which were rich in organelles, such as plastids, mitochondria, some sphaerosomes, surrounded by unit membrane and functioning as lysosome, and autophagic vacuole-like structure, some pro-Ubisch bodies and osmiophilic particles were distributed around in the cytoplasts (Plate X, 3). Moreover, there were some small vesicles surrounded by double-membrane and held together by endoplasmic reticulum. In the other sections, as we have reported, a lot of endoplasmic reticulam, well-defined and always parallel to each other, appeared in tapetum cells, some vacuoles with phagocytosis power were surrounded by endoplasmic reticulum and in addition, a number of simple plastids could be seen too (Plate X, 4).

After that, in fertile plant, the tapetum layer seemed to turn to degenerate. In some tapetal cells, bigger vacuoles formed, endoplasmic reticulum bent further, the Ubisch bodies migrated from the tapetum cell to the space near the cell wall (Plate X, 5). Other tapetum cells showed poorly defined structure, nucleus, simple plastids, little vacuoles and poorly defined endoplasmic reticulum could vaguely be seen; Ubisch bodies discharged from the tapetal cell could be seen too.

During the formation of exine, which was enriched with sporopollenin, and germination aperture of microspore, the walls between tapetal cells disappeared and cytoplasm almost disorganized. A large number of spherical structures, with different size and various electron density, some nuclear materials, Ubisch bodies and autophagic vacuoles appeared in this space (Plate X, 6).

2. Ultrastructure of tapetal cell in sterile plant

In sterile plant, the ultrastructure of tapetum cell was the same as in fertile form, at the stage of pollen mother cell. While its meiosis began, the ultrastructure of tapetum cell showed complicated changes and some significant difference from that of fertile form. They appeared as: the formation of concentric-circle shaped endoplasmic reticulum, and some vacuoles of various size near the endoplasmic reticulum (Plate XI, 1); Autophagic vacuoles (which were formed by an unit-membrane surrounded some organelles) and osmiophilic particles with irregular forms appeared (Plate XI, 2). Sometimes certain endoplasmic reticulum became parallel to each other and surrounded the vacuole in the tapetal cell. There were simple plastids and mitochondria too (Plate XI, 3). These pictures were similar to those shown in fertile plant tapetal cell at the period of tetrad formation.

Further degeneration of tapetum cell in sterile plant developed as follows: the cytoplasts disorganized gradually, vacuole-membrane dissolved and the main part of cell lost its structural features; only a little cytoplast remained near the cell margin, there were some degenerating organelles and formless endoplasmic reticulum could be seen in the cytoplasm (Plate XI, 4). Once the cytoplast in tapetum cell disorganized further and the cell became an empty space. In this space some nuclear material, sphere-shaped structures of all sizes, as well as autophagic vacuoles like structure could be seen (Plate XI, 5). When tapetum layer completely disorganized, the anther shrunk and the contents of tapetal cells dissolved entirely; only a few survivals could be seen. The wall between tapetal cells disappeared and left only a poorly definded limit between the tapetum layer and anther cavity. On the other hand, in the anthers which did not shrink, well-defind cell walls of tapetal cells could be seen clearly, but the protoplasts had been completely disorganized and only a little dissolved protoplasm remained (Plate XI, 6).

Discussion and Conclusion

A cytomorphological comparison between Taigu genic male-sterile wheat and nucleo-cytoplasmic interactive Timopheevi-type male sterile wheat (abbreviated as T-type male sterile wheat)

According to the records of cytomorphological observations on T-type male sterile wheat and in comparison with the present work, the distinctions between these two materials can be summarized as follows:

1. The critical period and final state of the abortion of microspores between Taigu genic male sterile wheat and T-type male-sterile wheat manifest significant difference.

In Taigu genic male-sterile wheat, abortion took place at the later period of microsporogenesis, i.e. from the formation of tetrad to the release of young microspores. After that, the anther development came to a stand-still, its contents began to disorganize and then a "non-pollen" type abortion was formed[3]. Abortion in T-type male sterile wheat occurred at a later period of microspore development (single nucleus pollen stage) and microspores failed to develop further into male gametophytes, becoming "empty pollen" at last[5].

2. The development and degeneration of tapetum cells in Taigu genic male sterile wheat and in T-type male sterile wheat are also different.

In the former, tapetum layer began to degeneration during the meiosis of pollen mother cell, while in T-type male sterile wheat the development of tapetum layer in male-sterile line was similar to that in the fertile line[5]. Sometimes the tapetal cells of sterile line accumulated less starch grains and existed longer than in fertile line[7]. This would suggest that the early disintegration of tapetum layer in Taigu genic male-sterile wheat plays a key role in the degeneration of microspores.

3. The chromosome behavior during the process of pollen mother cell meiosis in T-type male sterile wheat, as a whole, was normal, but some chromosome aberrations occurred in Taigu genic male sterile wheat

The real cause of chromosome aberration, whether genetic or enviromental (such as nutrient imbalance caused by tapetal cell's early degeneration), remaines unclear.

Such strong distinctions of cytomorphological feature between Taigu genic male sterile wheat and T-type male sterile wheat also prove that the genetic background of these materials are significantly different[1,2].

The ultrastructural characteristics of the development and degeneration of tapetum cell in Taigu genic male sterile wheat

We find that some electromicroscopic features of tapetum cells during the period of disorgnization are worthwhile to note. First, the activity of endoplasmic reticulum was rather exceptional. An example is the appearance of rich endoplasmic reticulum, whose membranes ran parallel each other or arranged in concentric circles in cross-section, and surrounded the vacuole with phagocytosis or some cytoplast. The changes of endoplasmic reticulum also could be found in fertile form, when tapetum cell development proceeded to "maturity" and then turned to degeneration. At this time, tapetal cells contained abundant and more extended endoplasmic reticulum. Perhaps, such ultrastructural features could be looked upon as a specific sign of tapetal cells before their disorganization.

Another feature was the formation of autophagic vacuoles and osmiophilic particles. In tapetal cells of sterile form, there were some autophagic vacuoles formed by a unit-membrane surrounding some organelles and some vacuoles with phagocytic ability which always contained concentric circle-shaped membrane structure or some vesicles and small particles. In fertile form, autophagic vacuoles and spharesomes were also formed; these were regarded as lysosomes during the later period of tapetum cell development. The appearance of such structures as described above indicates that the disorganization of tapetum layer is taking place. Moreover, during tapetum cell degeneration, specific osmiophilic particles appeared. This feature is similar to the report of De Vries (1970) regarding electromicroscopic observations on T-type male sterile wheat. In his report, the dark grains which occurred in T-type male sterile line were looked upon as the precusors of Ubisch body or were thought to contain some material that built Ubisch body[8]. We think that the appearance of osmiophilic particles may be considered as part of the degeneration of tapetum cells.

Thirdly, tapetum cells became vacuolized before their degeneration.

Other report on T-type male sterile wheat considered that the tapetum cells of the sterile and fertile (maintainer) line did not show any notable differences before the abortion; only in the later stage of microspore development vacuoles containing concentric membranous bodies were seen in the tapetal cells of maintainer line[6]. The ultrastructure of development and degeneration of tapetum cell in Taigu genic male sterile wheat were different from that of T-type male sterile wheat, and expressed distinct characteristics.

The ultrastructure of pollen mother cells in sterile form appeared significantly different from that of fertile form in Taigu genic male sterile wheat

Some concentric circle-shaped or bundle-shaped endoplasmic reticulum appeared in the pollen mother cells of sterile plant and when reduction division reached its diakinsis stage, the pollen mother cells showes significant vacuolization. These results are similar to those in T-type male sterile wheat examined by Turbin et al. (1975). They found at the period of reduction division prophase I, in pollen mother cell of T-type sterile line, endoplasmic reticulum sometimes appeared as a concentric circle-shaped structure, with ribosomes arranged densely on it, and also showed stronger vacuolization[9].

In addition, at various stage of microsporogenesis, some osmiophilic particles and sphere-shaped structures with different sizes and electron density could be found in the pollen mother cells. When such structures appeared, the microsporogenesis process departed from its normal path and aborted soon after.

At the early stage before its abortion, the pollen mother cells of Taigu sterile plant exhibited changes in fine structure. During the process of microsporogenesis changes sharpened and resulted in the complete degeneration of microspores. These facts favor our early view that there would be a process preceeding the disorganization of tapetum cells and microspores[4]. It has been reported that in T-type male sterile wheat, the cell organelles of both the male sterile and its maintainer lines appear essentially alike at the "shrinking" stage of microspore development and only at the "small vacuole" stage does the vacuolar system show significant difference between these two lines[6]. The fine structure in microsporogenesis and its abortion process were seen to be quite different between Taigu genic male sterile wheat and T-type male sterile wheat.

To sum up concerning the ultrastructural characters of tapetum layer and microsporogenesis in Taigu genic male sterile wheat, the conclusions we arrive at are as follows:

1. It seems advisable to pay even more attention to the activity of endoplasmic reticulum in the anther tissue of the sterile plant.

The significant characteristics is the formation of concentric circle-shaped endoplasmic reticulum. This structure could be found not only in tapetal cell but also in the process of microsporogenesis of sterile plant; even in T-type male sterile wheat could a similar structure be found. It is thought that the endoplasmic reticulum is closely related to the synthesis of protein. In this connection, it is suggested that in Taigu genic male sterile wheat the degeneration of anther tissues and microspores may be concerned with the change of synthesis of some proteins (inconcluding enzymes). It is necessary

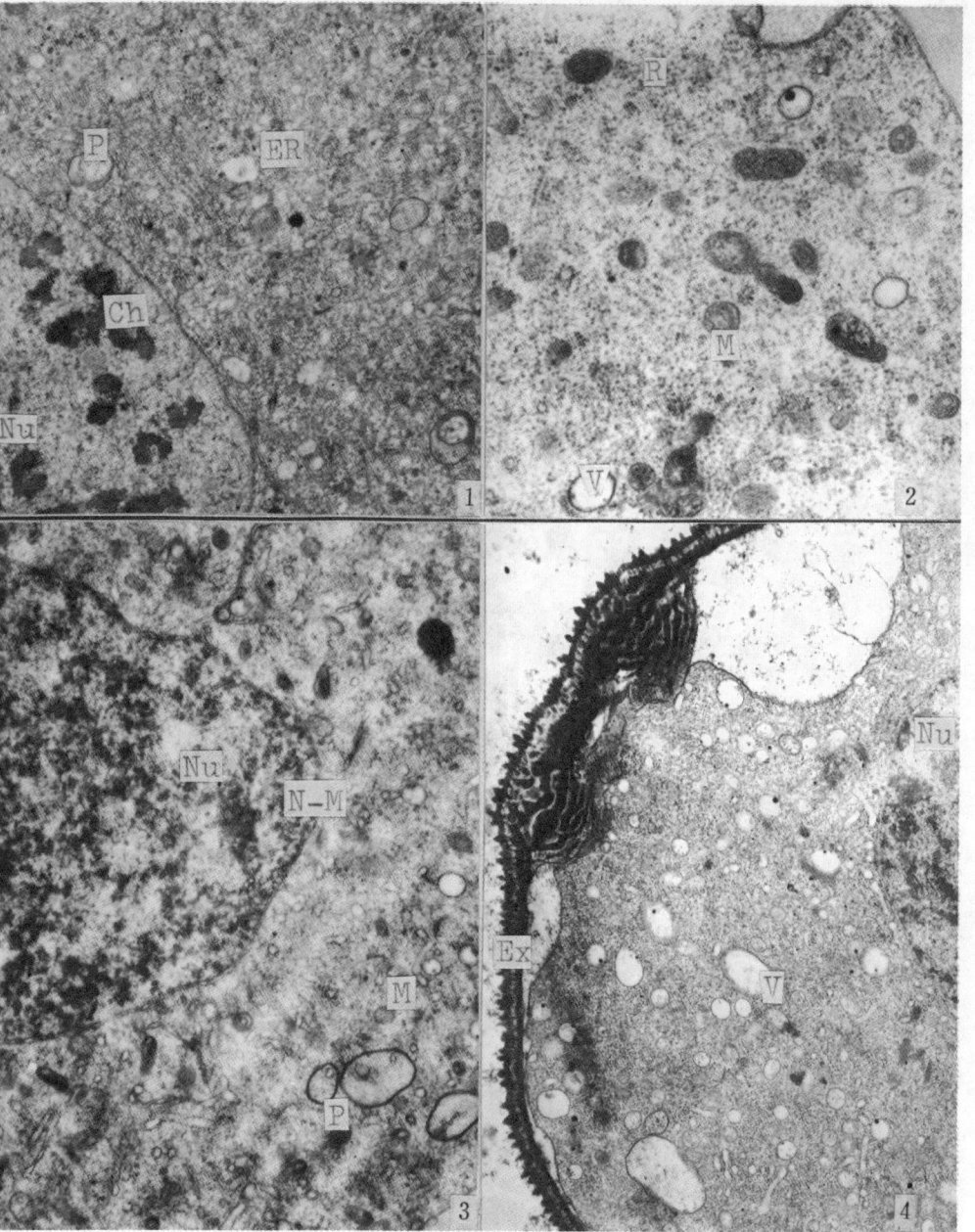

Plate VIII

1-4. All from the fertile form of Taigu genic male sterile wheat.
1. A part of the pollen mother cell at the period of Prophase I of reduction division. × 8800
2. A part of pollen mother cell during Anaphase I of meiosis. × 16000
3. A part of young microspore. × 11000
4. A part of microspore at the late stage of development. × 7000

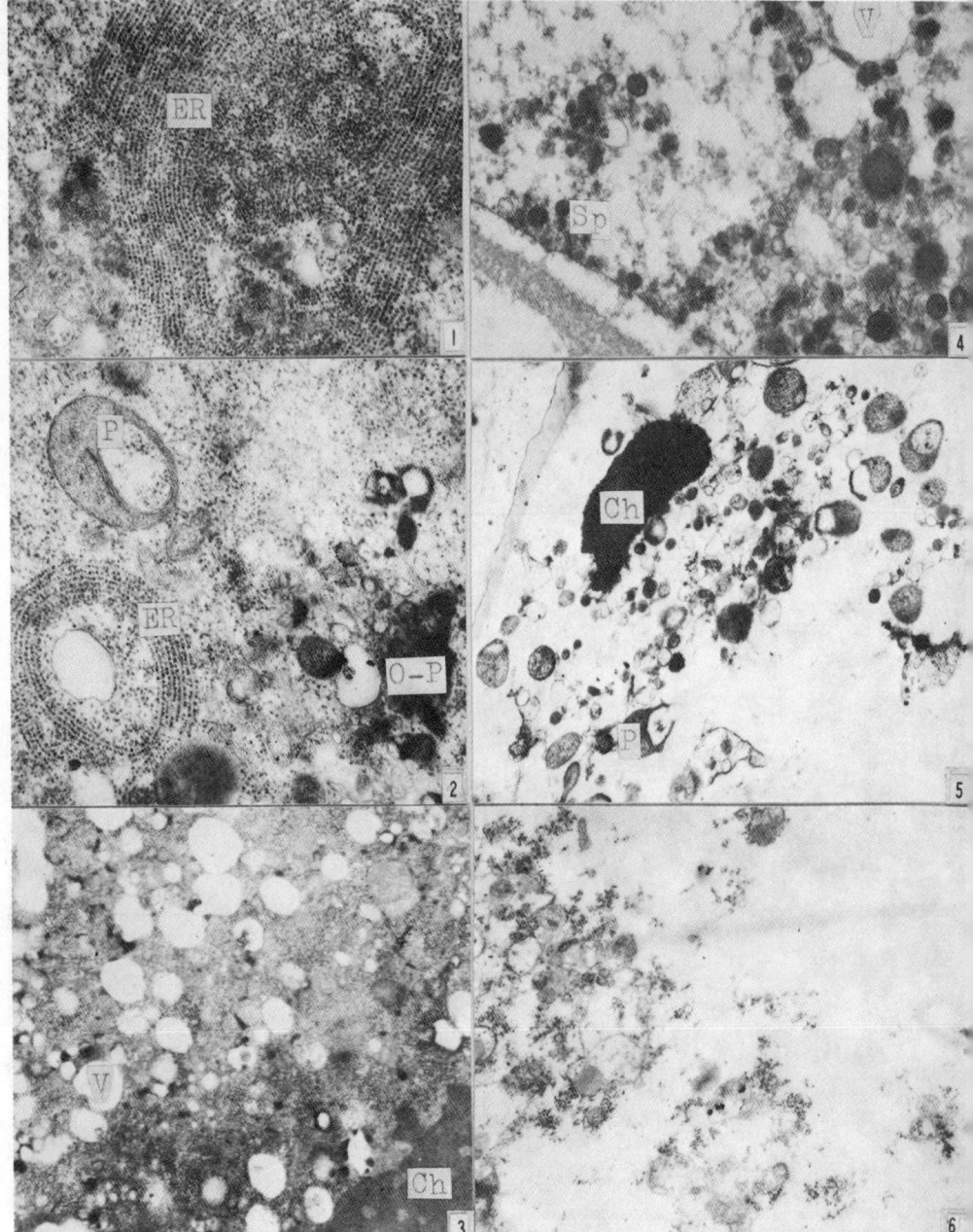

Plate IX

1-6. All are from the sterile form of Taigu genic male sterile wheat.
1. A part of P.M.C. showing a concentric circle-shaped endoplasmic reticulum. × 15000
2. A part of P.M.C. showing another concentric circle-shaped endoplasmic reticulum and some osmiophilic particles. × 15000
3. A part of P.M.C. at diakinesis, showing a lot of small vacuoles and some sphere-shaped structure. × 10400
4. A picture of degeneration microspore. × 11600
5. Another picture of degeneration microspore. × 7200
6. Still other picture of degeneration microspore. × 11000

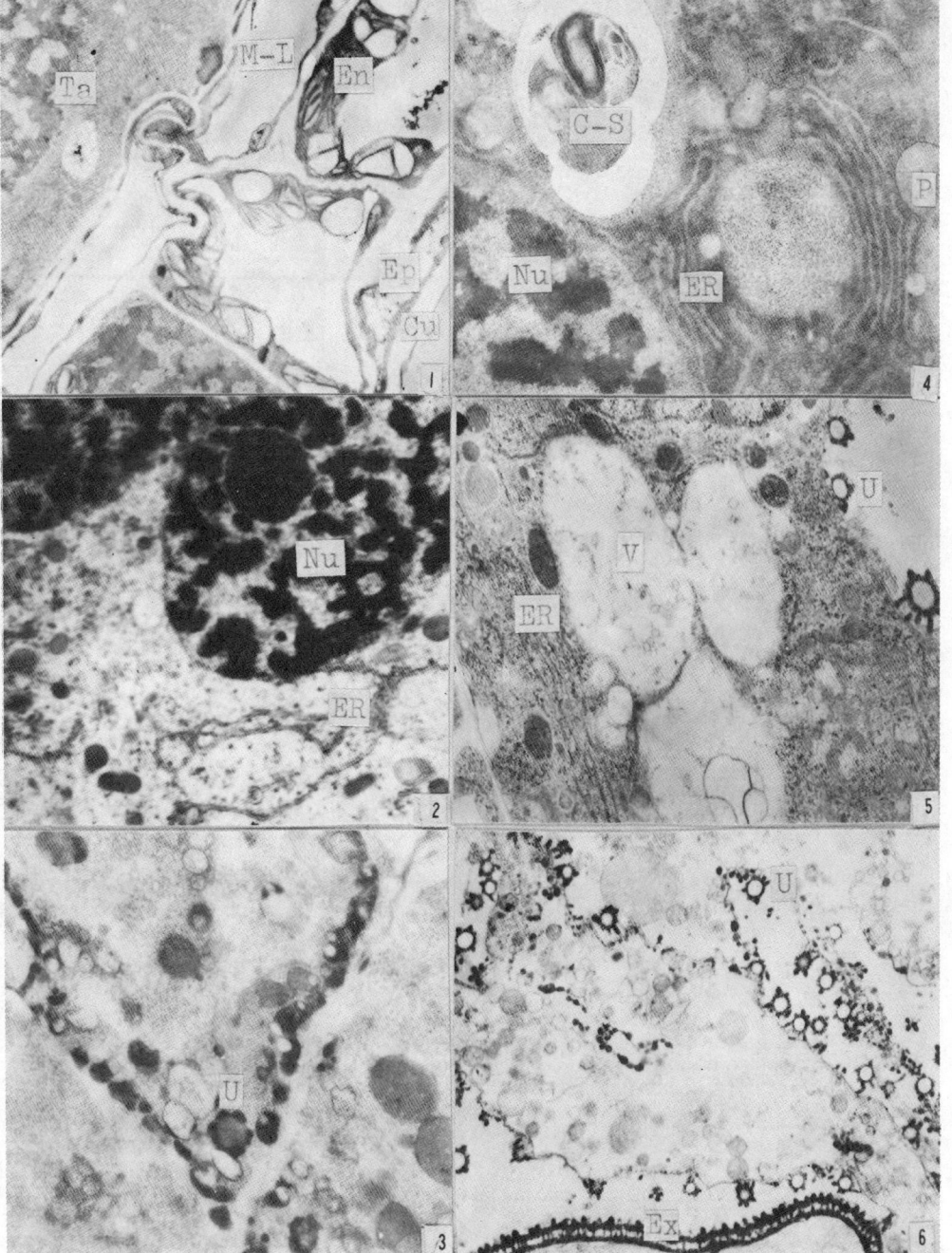

Plate X

1-6. All are from the fertile form of Taigu genic male sterile wheat.
1. A cross-section of anther wall layers during Prophase I of meiosis. ×3600
2. A part of tapetal cell at the period as in 1. ×7000
3. A part of tapetum cell at the period from the formation of tetrad to young microspore. ×9900
4. Another picture of tapetum cell at the period as in 3. ×9200
5. A part of tapetum cell at the development stage of microspore. ×7700
6. A part of tapetal cell at later stage of microspore development. ×6200

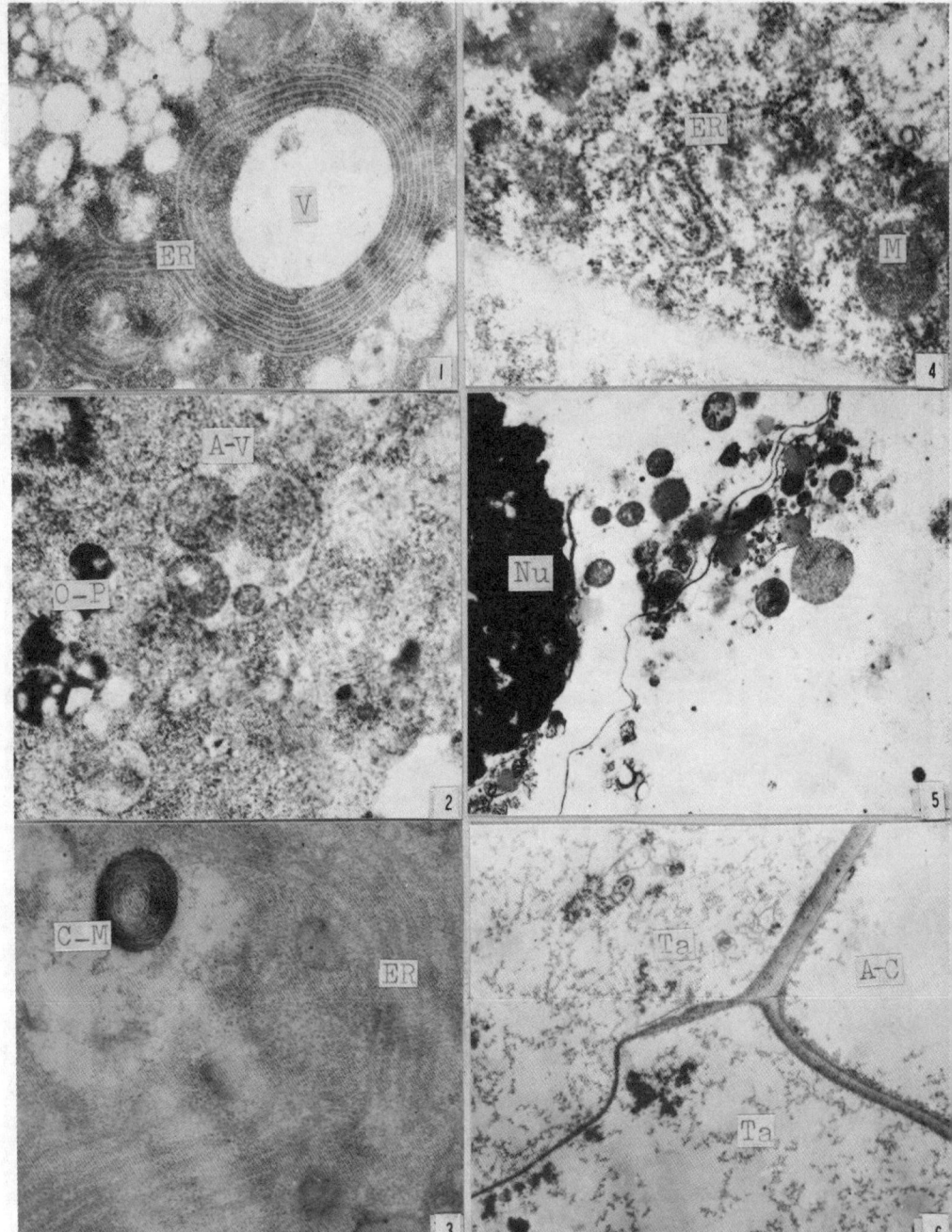

Plate XI

1-6. All are from the sterile form of Taigu genic male sterile wheat, and 1 – 3 are taken at the period of Prophase I of P.M.C.
1. A part of tapetal cell showing a concentric circle-shaped endoplasmic reticulum. × 18000
2. Showing autophagic vacuole and osmiophilic particles in tapetum cell. × 18000
3. A vacuole with phagocytosis ability was surrounded by the endoplasmic reticulum in tapetum cell. × 17500
4. A disorganization picture of tapetum cell. × 17500
5. Another disorganization picture of tapetum cell. × 7700
6. Still another disorganization picture of tapetum cell. × 5500

to make further research on the activities of protein and amino acid.

2. The formation of irregular osmiophilic particles may be looked upon as a sign of cell senescence

This material could be found in tapetum cells and microspores before their disorgnization. According to the electron density, these materials were similar to lipoid substances; the reasons for its formation are still unknown.

3. The formation of autophagic vacuoles with phagocytosis ability and spherosomes are regarded as the characteristic ultrastructural features in the cell degeneration

In addition, it seems that the increase of vacuolation either in tapetal cell or microspore signals final degeneration, the production of these structures preceeding the complete breakdown of protoplast.

References

[1] 邓景扬、高忠丽,1980,小麦显性雄性不育基因的发现与利用—太谷不育小麦鉴定总结,作物学报, 6（2）: 85-98。

[2] 王琳清、程俊源、施巾帼、高忠丽, 1980, 小麦显性单基因控制的雄性不育材料2-2-3的研究及其利用。中国农业科学,（2）: 1-8。

[3] 陈朱希昭, 1982, 太谷核不育小麦花药和花粉发育的细胞形态学观察。山西农业科学,（2）: 13-16。

[4] 陈朱希昭、陈耀堂、高信曾, 1983, 太谷核不育小麦花药组织和小孢子发育的超微结构研究（摘要）。《中国植物学会五十周年年会议文摘要汇编》第523页。

[5] 北京大学生物系植物遗传育种专业, 1976, 雄性不育系和雄性能育小麦花药组织粉发育细胞形态学观察。植物学报, 18（2）: 411-149。

[6] 胡适宜、王模善、徐丽云, 1977, 小麦雄性不育系和保持系的小孢子发育的电子显微镜研究。植物学报, 19（3）: 167-171。

[7] Joppa, L. R., F. H. McNeal, and J. R. Welsh, 1966, Pollen and anther development in cytoplasmic male sterile wheat (*T. aestivum* L.) *Crop Sci.,* **6**:296-297.

[8] De Vries, A. Ph., and J. S. Ie., 1970, Electron-microscopy on anther tissue and pollen of male-sterile and fertile wheat (*T. aestivum* L.) *Euphytica.,* **19**:103-120.

[9] Н. В. Турбин, А. Н. Палилова, 1975, Генети Ческие Основы Цитоплазматической Мужской Стерильности у Растений (101-113) *Наука и Техника,* Минск.

The Abnormal Change of Lysine-Rich Histone in the Pollen Mother Cell Nuclei before and after Meiosis

Xue Guangxing

(*Shanxi Academy of Agricultural Sciences*)

Deng Jingyang

(*Institute of Crop Breeding and Cultivation, Chinese Academy of Agricultural Sciences*)

Introduction

A chromosome is composed of DNA, basic protein histone and nonhistone. DNA-histone complex constitutes 70—90% of the total chromosome[6]. Histone has various molecular types. Phillips (1962) reported that it can be classified as lysine-rich histone (lysine: arginine >2) and arginine-rich histone (lysine: arginiee $\leqslant 1$) based on the lysine to arginine ratio.

Saffhill, et al. reported that H1 histone, one molecular type of lysine-rich histone, may conjugate at the fine fibre segment of chromatin subunits. The promoter region which is activated by RNA polymerase and is responsible for the synthesis of DNA may exist in the same segment[5]. Ceder et al. (1976) inferred that most probably H1 histone was involved in DNA transcription promoter region regulation. In recent years, encouraged by these achievements, more and more scientific workers have devoted themselves to the research on histone.

Great importance has been attached to the relationship between histone and male sterility[1,2,7,8]. On the different nucleoplasmic interaction male sterile materials, they all found that the histone level of the male sterile line is lower than that of the maintenance line and the restorer line. Dai et al.[1] reported the absence of one electrophoretic band from the free histone at the critical abortion moment, which established the relation between the composition of histone and the abortion.

Taigu genic male sterile wheat is the first mutant carrying a single dominant male sterile *Ta*1 gene[3]. In this paper, comparative observations on the decrease and increase of the lysine-rich histone around the onset of meiosis in PMCs were made between the sterile and fertile plants of the backcross progeny of Taigu wheat. Furthermore, the relation between the abnormal change of lysine-rich histone in sterile plant and the degeneration of PMCs was also studied. We hope it will be helpful to the understanding of the mechanism of male sterility.

Materials

The plants used in experiments were offsprings of Taigu male sterile wheat/4 × Baixiao-79-11 (*T. aestivum* L., 2n = 42) which were grown in 1983 in the nethouse of the Crop Institute, Chinese Academy of Agricultural Sciences with routine management. When most of the sterile and fertile plants could be identified from the floret anther shapes vigorous plants of similar phenotype were collected, then marked and fixed individually.

Methods

Plant samples were collected at field temperature 22–26°C, stored in ice-pot, fixed in 10% neutral formaldehyde (pH = 7.1), embedded in 46–48°C paraffin in the routine manner. Sections were cut at 10μ, stuck with Haupt's adhesive.

In our experiment the ammoniacal-silver method described by Black and Ansley[9], and the eosin-fast green staining method described by Bloch[10] were used with slight modifications.

In the whole process of ammoniacal-silver method redistilled water was used instead of distilled water. The sections underwent pretreatment for 3 hours in 10% neutral buffered formaldehyde, then stained in ammoniacal-silver solution for 45 seconds at temperature 25°C.

In the eosin-fast green method 0.07 mol/L Tris-HCl buffer was used as rinsing and differentiating solution. It is required that the concentration of eosin-Y and fast green should be strictly equal and should be twice as high as that used in Bloch method (pH = 8.15).The sections were stained for 3 hours and differentiated for 7 minutes. From hydrolyzation in trichloracetic acid to dehydration in pure ethanol, all processes were carried out in a refrigerator (0–4°C). Both reaction solutions were immediately prepared. All chemicals, except fast green (Fluka) were obtained from Beijing Chemical

Factory.

Control tests were as follows:

1. Materials collected were subjected to quick freezing and sectioning with cryostat microtome. The sections were immediately treated in 0.14% trypsin solution at 0–4°C for 2.5 hours, then fixed and dyed. With either ammoniacal-silver method or eosin-fast green method, nuclei were hardly ever stained.

2. Materials were extracted in $0.2N$ HCl at 0–4°C for an hour, then fixed, embedded, sectioned and dyed in the same jar together with the sample. With either staining method, the staining of control sections all became very weak due to the loss of most histone from extraction.

3. Adopting the method of Black[9], we treated the control sections before dyeing with nitrite solution which could block the ϵ-amine group of lysine. After this treatment, staining with ammoniacal-silver reagent resulted in a partial fading of the characteristic brownish yellow color of histone.

4. Following the description of Bloch[10], the control sections were extracted in 5 mol/L urea solution for 2 hours, then hydrolysed normaly and stained, the nuclear color of lysine did not appear any more.

Measurement of Absorption Curves

According to the criterion provided by Bennett[11], the pollen mother cell nuclei right at stage 2 of premeiotic interphase were selected from the ammoniacal-silver stained sections for spectrophotometric measurements.

Compared with the stage 1 nuclei, stage 2 interphase nuclei appeared to be more contracted, homogeneous denser and the diffuse reticulate appearance was absent (Plate XII 1 and 2). Compared with stage 3 nuclei, the outlines of stage 2 nuclei were more round, clear and uniform. The faintly visible nucleoli did not project the surface of nucleus (Plate XII, 2 and 3).

Spectrophotometric measurements were made in accordance with the Nougarède[12] method on the instrument of MPV-2 model. The wavelength used to measure absorption curve was 400–750 nm. All individual nuclei were subjected to 30 replicated measurements at each wavelength. 15 nuclei were measured from five florets of three different sterile or fertile plants. The absorption curve was drawn on the mean extinction values[1] of 15 nuclei and their standard deviations.

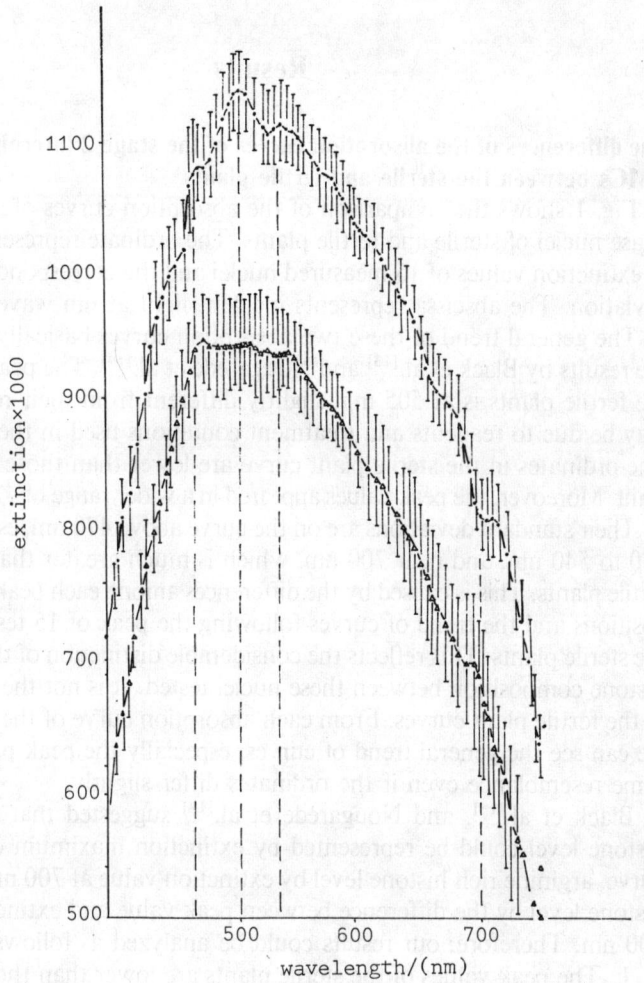

Fig.1 Absorption spectra of stage 2 premeiotic interphase nuclei in the genic male sterile wheat plants and their normal fertile plants. Chromatin-stained with ammoniacal-silver; extinction measured by the single-wavelength method.
• male-fertile; △ male-sterile.

---

1) extinction $= 1,000 \log \dfrac{\text{blank optical density}}{\text{sample optical density}}$

Results

The differences of the absorption curves of the stage 2 interphase nuclei of PMCs between the sterile and fertile plants

Fig. 1 shows the comparison of the absorption curves of stage 2 interphase nuclei of sterile and fertile plants. The ordinate represents the mean of extinction values of 15 measured nuclei and the corresponding standard deviation. The abscissa represents incident light at nm wavelength.

The general trend of these two absorption curves basically agreed with the results by Black et al.[13] and Nougarède et al.[14]. The peak value from the fertile plants is at 505 mμ, slightly different from their results, which may be due to reagents and treatment conditions used in the experiment. The ordinates in the sterile plant curve are lower than those of the fertile plant. Moreover, the peak values appeared in a wider range of 470 to 540 nm.

Their standard deviations are on the curve above 470 nm, especially from 470 to 540 nm, and near 700 nm, which is much greater than that of the fertile plants. This is caused by the differences among each peak values, peak positions and the trend of curves following the peak of 15 tested nuclei of the sterile plants. This reflects the considerable distinction of the chromatin histone composition between these nuclei tested. It is not the same as that in the fertile plant curves. From each absorption curve of the tested nuclei we can see the general trend of curves, especially the peak position, bears some resemblance even if the ordinates differ slightly.

Black et al.[15]. and Nougarède et al.[14] suggested that nuclear total histone level could be represented by extinction maximum of absorption curve, arginine-rich histone level by extinction value at 700 nm, lysine-rich histone level by the difference between peak value and extinction value at 700 nm. Therefore, our results could be analyzed as follows.

1. The peak values of the sterile plants are lower than those of the fertile plants, indicating that as far as the mean value is concerned, the total histone level in the stage 2 of PMC nuclei is lower in sterile plants than in fertile plants.

2. As for the sterile plants, the shape of the absorption curve peak in the range of 470–540 nm was displayed smoothly and widely. Each peak position differed significantly from each other but the differences in peak value were very small, indicating that these measured nuclei were similar in total histone level. Moreover, the standard deviation of the absorption curve around 700 nm was very significant, showing that the tested nuclei

had great difference in arginine-rich histone content. Thus the difference between the peak value and the extinction value at 700 nm, i.e. lysine-rich histone content also showed great heterogeneity among stage 2 interphase PMC nuclei of the tested sterile plants. This result will be confirmed by the eosin-fast green staining described below.

Eosin-fast green staining of PMC nuclei at stage 2 and 3 of premeiotic interphase

Bloch[10] established that when a group of cells could be divided into two types: lysine-rich cells and arginine-rich ones, after stained with fast green and eosin-Y in combination, the former nuclei exclusively bound eosin, exhibiting a bright pink. and the latter ones selectively bound fast green, stained green color. If certain nuclei contained these two types of histones simultaneously and both could undergo positive reaction respectively, nuclei would show dark purple color. The occurrence of eosinophilic positive reaction suggested the existance of lysine-rich histone. Someone[12] had made a cytochemical quantitative assay of the two types of histone with this staining method.

After eosin-fast green staining, PMC nuclei at stage 2 of premeiotic interphase for fertile plant showed dark purple color (Plate XII, 5). At stage 3 eosinophilia of these nuclei decreased a little. The nuclei color of the sterile plants at the stage 2 and stage 3 varied among blue-green, dark blue and dark purple colors (Plate-XII, 11–13). Its eosinophilia existed observable distinction, i.e. lysine-rich histone content varied among different nuclei. This agreed with the absorption curves obtained by the ammoniacal-silver method.

In the test, prior to onset of meiosis cohering or decomposing cells were observed among the PMCs of sterile plant. However, not only the cells selected for microspectrophotometric measurements after stining by ammoniacal-silver method, but also the ones used for color comparision or photographing were all morphologically normal cells. Although these PMCs were entirely normal in appearance, in fact, for the histone component and the level of lysine-rich histone content of their nuclear chromatin there existed quite obvious diversity among different cells of sterile plant and between the nuclei of sterile and fertile plants at the same stage. The difference might be one of the early signs of abnormal meiosis, degeneration and decomposition of PMCs, and heterogeneous generation and development of subsequent microspores of the sterile plant.

Before and after pachytene, the distinction of eosinophilia became more evident

Eosinophilia of PMC nuclei for fertile plant tended to decrease after the stage 2 of premeiotic interphase (Plate XII, 6-9). At diplotene phase eosinophilic reaction could hardly be observed. After diakinesis their nuclei all stained green.

In contrast to the fertile plant, eosinophilia of PMC nuclei in sterile plant didn't tend to decrease after the stage 2 even after the start of meiosis, rather the following complicated circumstances appeared.

1. Eosinophilia of some nuclei at leptotene for sterile plant was obviously smaller than that of fertile plant and even became negative reaction nuclei showing green color, while some nuclei had much stronger eosinophilia than that of fertile plant (Plate XII, 14).

2. In the test for sterile plants most PMCs observed stayed in pachytene, which was different from the fertile plants. For the sterile plant about 30% of pachytene nuclei were highly eosinophilic, 10% displayed green color and showing eosin negative reaction and the rest were mostly intermediate between the above two extremes; among which more than one half had higher eosinophilia than nuclei of fertile plants at the same phase (Plate XII, 15-16).

3. The diakinesis phase was observed in five anthers of sterile plants and all were highly eosinophilic, regardless of the variation in degree (Plate XII, 17).

4. With the eosin-fast green staining, metaphase I and subsequent phases of sterile plant were similar to that of fertile plant, exhibiting green color (Plate XII, 10). After this phase, degeneration or decomposition of PMCs in sterile plant accelerated abruptly and asynchronous meiosis was very common. In the same anther sac, pachytene, diakinesis, metaphase I, dyads, microspores could be observed at the same time, metaphase I were few, and mingled among degenerated or decomposed and nondecomposed PMCs existing in different stage of meiosis.

Overall, after the onset of meiosis and espesially before and after pachytene, diversity in eosinophilia, i.e. differences in lysine-rich histone content among the PMC nuclei of the sterile plant and between sterile and fertile plants became more significant than that before the onset of meiosis.

Discussion

DNA required for the mitosis of somatic cells is synthesized in its interphase. For PMCs in their interphase, only the large part of DNA required for meiosis is synthesized. The complemental component should be completed in zygotene-pachytene[6]. Interphase-pachytene are also important phases for protein synthesis.

Based on the study of meristem of pea axillary buds before and after topping, Nougarède[12] pointed out that after release of inhibition on the axillary buds, lysine-rich histone content in its interphase nuclei increased with the increase of cell mitotic activity.

In our present work, we observed that when the eosin-fast green method was used to stain nuclear histones, sporogenous cells of the fertile plants stained green. When they have transformed into PMC interphase nuclei, eosinophilic substance started to appear and increased subsequently, meaning that lysine-rich histone content increased pronouncedly (Plate XII, 4). After the onset of meiosis, eosinophilia of chromation of PMC nuclei was decreased sharply, it became too slight to be observed till diakinesis phase (Plate XII, 6-9). Eosinophilic substance only occurred in nuclear chromatin in an interval between interphase and early diakinesis.

From the above facts, one may notice that the increase (or occurrence) of lysine-rich histone content is related with the active DNA synthesis. Therefore the question may be raised whether this kind of histone has important influence on the process of meiosis, whether its content is proper in the chromotin of PMC nuclei. Our present study, using Taigu male sterile wheat. would seem to indicate that this is so.

According to the concept of cell cycle, PMC nuclei at stage 2 of premeiotic interphase should be at the latter half of G_1 phase and just at an important moment before DNA synthesis phase[11]. At this time their DNA content is 2c.

Just at this phase, by means of the two different staining methods, we compared sterile and fertile plants. Difference was observed in extinctions of their peaks for absorption curves, also differentiation in peak position. In addition to qualitative differentiation in absorption curves, at the same time varied lysine-rich histone content was also observed among nuclei at stage 2 of premeiotic interphase for sterile plant. The two methods obtained basically similar results.

Abnormal levels of lysine-rich histone in chromatin of stage 2 can easily interfere with DNA and protein synthesis during stage 3. After onset of meiosis, especially before and after pachytene, diversity in lysine-rich histone content became more significant among PMC nuclei of sterile plant and between sterile and fertile plants. This should be one of the direct consequences of above influence, which at the same time will interrupt DNA or protein synthesis during zygotene-pachytene, so as to interfere with autosynapsis, and bring more important influence on subsequent procedure of meiosis.

Chen, Zhu, et al.[4] had found the endoplasmic reticulum and ribosome of sterile plant differed from that of fertile plant around the first prophase, and abortion accelerated after diakinesis. In the present work, we also observed nonsynchronization of meiosis, degeneration and decomposition of PMCs of sterile plants accelerated after first metaphase, which was consistent with the above observation.

However, before and after onset of meiosis the histones in chromatin of PMC nuclei for sterile plant, especially the abnormality of lysine-rich histone content, should not be the primary effect of $Ta1$ gene expression, nor the only mechanism of their abortion.

Although abortion was accelerated after metaphase I quite a few PMCs narrowly completed meiosis and formed microspores. This is due to the variation in the content of lysine-rich histone in the nuclei at stage 2 and stage 3 of premeiotic interphase of sterile plant, some nuclei have no obvious difference from that of fertile plant at the same phase. In the subsequent steps of meiosis, a certain number of PMCs has appearance similar to the nuclei of fertile plant at the same phase. Their abortion could be caused by other mechanisms.

References

[1] 戴尧仁等，1978，水稻二九南一号雄性不育系及相应保持系花药中某些呼吸酶和游离组蛋白的比较研究，遗传学报，5（3）：227—234。
[2] 中国科学院遗传研究所四室二组，1976，细胞质雄性不育高粱及可育相似体的细胞化学初研究，6（2）：156—158。
[3] 邓景扬、高忠丽，1980，小麦显性雄性不育基因的发现与利用—太谷不育小麦鉴定总结，作物学报，6（2）：85—98。
[4] 陈朱希昭等，1984，太谷核不育小麦花药组织和小孢子发生的超微结构研究，植物学报，26（3）：235—240。
[5] 孙毓麟，1978，关于真核细胞染色质亚单位的研究，科学通报，（2）：78—87。
[6] 伊藤道夫，1975，减数分裂，东京大学出版会出版（王瑞丰译，1979，科学出版社）。

[7] Alam, s. and P. C. Sandal, 1967, Cyto-histological investigation of pollen abortion in male-sterile sudangrass, *Crop Sci.,* **7**: 587–589.

[8] Chauhan, S. V. S. and T. Kinoshita, 1979, Histochemical localization of histones, DNA and proteins in the anthers of male-fertile and male-sterile plants, **29**(4): 287–293.

[9] Black, M. M. and H. R. Ansley, 1964, Histone staining with ammoniacal silver, *Science,* **143**: 693–695.

[10] Bloch, D. P., 1966, Histone differentiation and nuclear activity, *Chromosoma* (Berl.), **19**: 317–339.

[11] Bennett, M. D. et al., 1973, Cell development in the anther, the ovule, and the young seed of *Triticum aestivum* L. var. chinese spring. Philosophical Transactions of the Royal society of London, serie B, **226**: 39–81.

[12] Nougarède, A. et P. Rondet, 1977, Les histones nucléaires et leurs variations dans les méristèmes axillaires du pois (Pisum Sativum L., var. nain hâtif d'Annonay), soumis ou non à la dominance apicale. C. R. Acad. Sci. Paris, t. **287**, série D: 623–626.

[13] Black, M. M. and H. R. Ansley, 1965(b), Antigen-induced changes in lymphoid cell histones. II. Reginal lymph nodes. *J. Cell biol.,* **26**: 797–803.

[14] Nougarède, A. et Rondet, P., 1975, Les protéines nucléaires basiques des bourgeons axillaires et terminaux du Pisum Sativum L. XII Intern. Bot. Congr. Leningrad(U.S.S.R): pp189–221.

[15] Black, M. M. and Ansley, H. R., 1966, Histone specificity revealed by ammoniacal silver staining. *J. Histochem. Cytochem.,* **14**: 177–181.

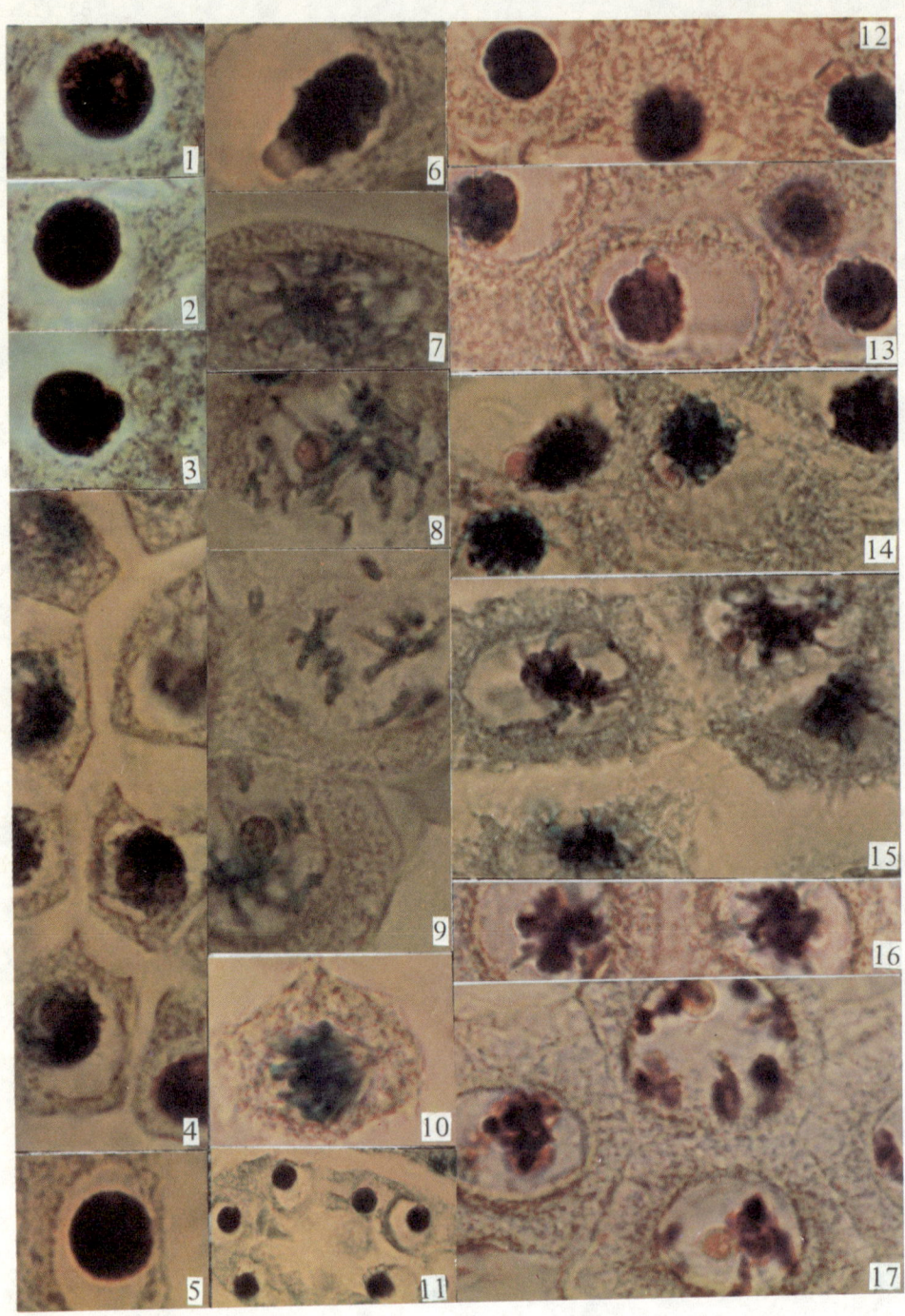

Plate XII

1-3. Nuclear chromatin stained by the ammoniacal-silver method according to the description of Black and Ansley, from male-fertile Taigu wheat plant ($\times$ 2200).
1. Stage-1 of premeiotic interphase P.M.C. nucleus;
2. Stage-2 of premeiotic interphase P.M.C. nucleus;
3. Stage-3 of premeiotic interphase P.M.C. nucleus.
4-17. Nuclear chromatin stained by the eosin-fast green method on the basis of the description of Bloch. 4-10, from male-fertile Taigu wheat plant (4-9; $\times$ 2200, 10: $\times$ 1320).
11-17 From male-sterile Taigu wheat plant (11: $\times$ 528, 12-17: $\times$ 1320).
4. Sporogenous cell nuclei transforming to interphase P.M.C. nuclei;
5. Stage-2 of premeiotic interphase P.M.C. nucleus;
6. Leptotene phase;
7. Early pachytene phase;
8. Green early diplotene phase;
9. Green early diakinesis phase;
10. Green metaphase-I phase.
10-13 Heterogeneity in chromatin eosinophilia of P.M.C. interphase nuclei.
14-16 Significant diversity in chromatin eosinophilia of P.M.C. nuclei at leptotene and pachytene phase.
17. Highly eosinophlic diakinesis nuclei.

COMPARATIVE ANALYSIS OF FREE AMINO ACIDS IN STERILE AND FERTILE ANTHERS

Zhu Guanglian   T. H. Tsao

(*Department of Biology, Peking University*)

Natural mutants in which male sterility is controlled by dominant gene of the nucleus have been reported only in a few cases. Correspondingly, there has been very limited research work done on the biochemistry and physiology of genic male sterility. However, work done on cytoplasmic male sterile plants has shown that the free proline content of the anthers is closely related to male sterility. In both cytoplasmic male sterile anthers and normal fertile anthers, at the stage of reduction division of microspore mother cells, the content of free proline is low, and there is little difference between the two kinds of anthers. After reduction division of the microspore mother cells, the content of free proline in normal fertile anthers rises abruptly, while that of the male sterile anthers remains low or even absent. This phenomenon was first reported in wheat and corn[10] and later confirmed by other workers in corn[8,12,18], sorghum[6,11], rice[2,15], cabbage, onions and other vegetable crops[9] and wheat[3]. Furthermore, Brooks (1962), Kern and Atkins[11] (1972) reported that in sorghum the normal fertile anthers contained more alanine than the male sterile anthers, but Khoo and Stinson (1957), Duvick (1965) obtained opposite results in corn. There are also reports on the difference in other free amino acids, for instance, asparagine was reported to be higher in male sterile wheat anthers[8,10], and the male sterile sorghum anthers were reported to contain more glycine. Most of the work mentioned above was done in the 1960's and the method adopted was mainly paper chromatography which has limitaitons for quantitative evaluation.

To the knowledge of the authors, there has been no report on the difference in amino acid composition with regard to genic male sterility. The present investigation was undertaken to make a comparative analysis of the free amino acid composition in anthers of both genic male-sterile and normal Taigu wheat plants with the aim of getting some information on the biochemical processes leading to pollen abortion. The free amino acid composition of the pistils of both types of plants was also analyzed for comparison.

Material and Methods

Seeds from open-pollinated Taigu genic male sterile plants were sown in the greenhouse where the temperature varied from 25 to 30°C in the day time, and from 15 to 20°C at night. As expected, the plants segregated as male sterile and normal fertile in a 1:1 ratio, that is, about half of the wheat plants was male sterile, the other half being normal fertile. As soon as the sterility or fertility of the main spike was recognizable from the characteristic appearance, the spikes of the tillers of the same plant were removed. Anthers from the middle parts of the spikes of the tillers were squashed and stained with acetocarmine or carbofuchsin to determine the developmental stages of microspore within the anthers. Samples from the middle parts of the tiller spikes of more than 10 plants from both sterile and fertile material were analyzed. The pistils of the corresponding material were also sampled for analysis. The fresh weight of each sample was about 20 mg. consisting of about 30-150 anthers or about the same number of pistils. The dry weights of anthers or pistils were obtained from parallel specimens. When we say the binucleate stage of the sterile anthers, we really mean the stage corresponding to the binucleate stage of anthers of normal plants, since at that time the anthers of the sterile plants have become empty. This stage was ascertained according to the length of the spike and the distance from the flag leaf to the leaf ear of the leaf next to it.

Extraction and analysis of free amino acids

For extraction the method of Barnent (1966) and Kern and Atkins (1972) was employed with modification. The anther and pistil samples were ground in a small glass homogenizer and then extracted 4 times, each with 2 ml 80% ethyl alcohol. The extracts were combined, and ethyl alcohol was evaporated on a water bath at 80°C. The amino acid extract was reduced to a volume of 0.5 ml on a boiling water bath., 5% sulfo-salicylic acid was added in the proportion of 1:1. Centrifuge for 15 min. at $15,000 \times g$. The supernatant was analyzed for amino acids, using an automatic amino acid analyzer, model Beckman 121-BM.

Experimental Results
Analysis of the free amino acids in anthers

The results obtained with anthers of 3 developmental stages are presented in Table 1. Data show that at R.D. of the microspore mother cells, there was no significant difference between the content of free amino acids in male sterile and normal anthers, except that the content of alanine seemed to be lower in the fertile anthers than in the sterile ones. At the stage of release of microspores, the amino acid composition of both types of anthers underwent remarkable change; most noticeable changes were observed in proline, alanine, lysine and arginine.

1, Changes in proline

Table 1. Contents of free amino acids in the anthers of genic male sterile wheat plants and their normal fertile plants (μg AA/mg dried anther)

developmental stage amino acid	R.D. stage			release of microspores			binucleate pollen		
	F[1]	S[2]	F/S	F	S	F/S	F	S	F/S
Asp+Asn	1.07	1.10	0.97	0.90	0.60	1.50	1.06	1.41	0.75
Thr	0.58	0.80	0.73	0.77	1.33	0.58	1.06	1.38	0.77
Ser	0.97	0.80	1.21	2.02	3.28	0.62	1.98	3.93	0.50
Glu	1.94	2.00	0.97	2.57	1.38	1.86	2.80	5.34	0.52
Pro	0.77	0.72	1.07	2.45	0.24	10.21	9.40	0.99	9.49
Gly	0.66	0.62	1.06	0.48	1.43	0.33	0.56	1.59	0.35
Ala	1.23	2.75	0.45	4.73	2.76	1.71	7.20	2.94	2.45
CysH	0.10	—		—	0.24		—	0.09	
Val	1.34	0.82	1.63	0.91	1.81	0.50	1.80	2.07	0.87
Met	0.18	0.30	0.60	0.21	0.38	0.55	0.54	0.24	2.25
Ileu	0.55	0.35	1.57	0.34	0.81	0.42	0.65	0.93	0.70
Leu	0.74	0.54	1.37	0.53	1.19	0.46	0.88	1.38	0.64
Tyr	0.45	0.62	0.73	0.50	0.57	0.88	0.52	0.95	0.55
Phe	0.57	0.60	0.95	0.53	0.52	1.02	0.88	1.41	0.63
Lys	1.43	1.49	0.96	1.51	0.24	6.29	1.56	2.58	0.61
His	0.44	0.38	1.16	0.38	0.71	0.54	0.34	0.84	0.41
Try	0.17	0.27	0.63	0.12	0.47	0.26	0.14	0.36	0.39
Arg	1.00	0.88	1.14	2.13	1.14	1.87	1.54	2.64	0.58
total	14.2	15.0	0.95	21.1	19.1	1.10	32.9	31.2	1.05

1) fertile plants
2) sterile plants

Table 2. Contents of free amino acids in the pistils of both types of plants when the pollen was at the stage of release of microspores (μg/mg. dried pistil)

amino acid	F[1]	S[2]	F/S
Asp.+Asn.	1.94	1.42	1.32
Thr.	1.03	1.33	0.77
Ser.	2.15	2.58	0.83
Glu.	1.74	5.91	0.29
Pro.	–	–	
Gly.	0.87	0.49	1.78
Ala.	3.60	2.49	1.45
CysH.	–	0.04	
Val.	0.79	0.98	0.81
Met.	0.18	0.12	1.50
Ileu.	0.78	0.94	0.83
Leu.	0.54	0.45	1.20
Tyro.	1.32	1.73	0.76
Phe.	0.54	1.02	0.53
Lys.	2.20	1.02	2.16
His.	0.33	0.53	0.62
Try.	0.20	0.16	1.25
Arg.	1.05	1.07	0.96
total	19.2	21.8	0.88

1) fertile plants
2) sterile plants

From R. D. to release of microspores, the content of proline in normal fertile anthers rose from 0.77 µg/mg·dry wt. to 2.45, while that in nuclear sterile anthers, it declined from 0.72 to 0.24, resulting in a 10 fold difference between the two types of anthers. It is amazing to note that in anthers with binucleate pollen, the content of free proline continued to increase to 9.4 µg, constituting almost 30% of the total free amino acids of the anther, amounting to 1% of the dry material of the anther. At the same time, the proline content of the male sterile anthers, although increased a little, only amounted to 1/10 of that in normal fertile anthers.

2. Changes in alanine

At the stage of release of microspores, the content of alanine was 4.73 µg/mg dry anther. It continued to increase, reaching 7.2 µg at binucleate stage, constituting the next highest amino acid in the anther, only lower than proline.While alanine increased greatly in fertile anthers, it remained relatively constant in sterile anthers. But even at binucleate stage, alanine content in fertile anthers was only 2.45 times of that in sterile anthers, a difference less conspicuous than proline.

3. Changes in lysine and arginine

The change of the basic amino acids lysine and arginine deserves attention. It can be seen that lysine and arginine, especially lysine, remained relatively constant in fertile anthers throughout all the three developmental stages,while they underwent a greater change in sterile anthers. A lower content of these two basic amino acids was observed in sterile anthers at microspore releasing stage when the microspores were degenerating repidly. At the early binucleate stge of pollen, however, the content of these two basic amino acids was higher in sterile anthers.

The total free amino acids between normal fertile and male sterile anthers did not undergo much change, as showin in the last line of Table 1. It can also be noticed that with the advance of development, the total free amiro acids of both types of anthers increased.

Analysis of free amino acids in pistils

The difference in the amount of glutamic acid between pistils of sterile and fertile plants at microspore releasing stage is remarkable. In pistils of male sterile flowers, glutamic acid reached a high amount of 5.91 µg/mg, more than three times of that in pistils of fetile flowes. Since glutamic acid is the precursor of proline synthesis, and microspore releasing stage is the stage when proline accumulates rapidly in fertile anthers, it seems likely that the accumulation of proline in fertile anthers leads to the shortage of its precursor, glutamic acid, in the pistils. In other words, the incapability of proline accumulation in sterile anthers is probably the cause of its abundance in pistils. This result seems to indicate that although the expression of *Ta*1 gene is mainly confined to anthers, it may influence the composition of the pistils to some extent (Table 2).

It is noticeable that while proline constituted the highest amino acid in anthers, and it showed a tenfold difference between fertile and sterile anthers, it could not be detected in pistils at the microspore releasing stage.

Our result concerning the lack of free proline in pistils compares well with the results of Fujushita and Britikov and Musatova[9,5]. Fujnushita showed that no proline was detectable in pistils of either fertile or sterile materials in a number of vegetable crops. Britikov and Musatova determined the free proline content of the pollen, pistils and sometimes the vegetative organs of approximately 200 plant species. In all cases the proline content of the pistils ws lower than that of the pollen.

Discussion

Results presented in this paper concerning the high content of free proline in fertile anthers and the remarkable 10 fold difference between the sterile and the fertile anthers lead us to consider that the same metabolic pecularities of proline occur both in nuclear sterile anthers and cytoplasmic sterile anthers. In a previous paper we have determined the proline content in anthers of both sterile and fertile Taigu wheat plants growing under field conditions by colorimetric methods. The same results were obtained. Under field conditions, the proline content of the normal fertile anthers was even higher, reaching 1.6% of the dry weight of the anther, and consituting 50% of the total free amino acids.

Pàlfi (1981) put forth the viewpoint that the amount of free proline in anthers can be taken as an index of sterility or fertility. As to the physiological role of free proline, Tupy (1964) found that free proline is an important energy and nitrogen source[19,20]. According to Linskens and Schrauwen, Dashek, and Zhang, free proline is directly utilized in protein synthesis during pollen germination and pollen tube growth[7,13,14,21]. The results of the present paper lead us to think that free proline may have similar physiological function in the early development of the pollen grains, its absence may accelerate or directly lead to the abortion of microspores.

The remarkable difference in the two basic amino acids, i.e., lysine and arginine at microspore releasing stage and in early binucleate stage between sterile and fertile anthers has not been reported or emphasized in the literature.

Lysine and arginine are the main constituents of histone, which is a chromatin protein existing in the DNA spirals. As is well known, histone inhibits the transcription of DNA into RNA, thus regulating the expression of genes. It seems that the difference in lysine and arginine between the two types of anthers is of primary importance to the abortion of the sterile anthers.

Our result about the difference in alanine is consistent with that of Brooks and Kern and Atkins in sorghum[6,11], but conflicts with that of Khoo and Stinson and Duvick in corn[8,12]. The accumulation of asparagine in sterile anthers as reported by Fukasawa in wheat and by Duvick in corn could not be repeated by us with genic male sterils material[8,10].

As is known, nucleo-cytoplasmic male sterility, studied by early investigators and genic male sterility, reported in the present paper, are controlled by entirely different mechanisms. The latter is controlled by the nucleus, and the former is controlled by mitochondrial DNA, in addition to being influenced by the nucleus. In spite of this different genetic background, at the critical period of pollen abortion, there appeared similar difference in the composition of amino acids, such as in proline, in alanine and glycine, in sterile and fertile anthers of Taigu wheat. It is therefore logical to think that changes in amino acid composition in sterile anthers is not the primary reaction of gene expression. It may be one of the accompanying phenomena resulted from metabolic disorders, a view early held by Britikov[5]. However, this does not rule out the possibility that the lack of proline, a physiologically active amino acid, could accelerate pollen abortion.

Summary

Comparative analyses were made on the composition of free amino acids in anthers of Taigu genic male sterile wheat plants and their normal fertile counterparts. An automatic amino acid analyzer of Beckman 121-BM model was used. Results showed that at reduction division of microspore mother cells, no significant difference could be detected between the free amino acid composition of both types of anthers, but during the release of tetrads and at the early binucleate stage of the pollen, the composition of free amino acids underwent remarkable changes. The change of proline is most pronounced. Proline content in fertile anthers at the microspore releasing stage and at the binucleate stage was 10 times higher than that in sterile anthers. Alanine had the same tendency of change with proline, but the difference was less conspicuous than proline. Lysine and arginine, especially lysine, remained relatively constant in fertile anthers, while they underwent a greater change in sterile anthers. A lower content of these two basic amino acids was observed in sterile anthers at the microspore releasing stage when the microspores were degenerating rapidly. At the early binucleate stage of pollen, the content of these two amino acids became higher in sterile anthers.

The observed differences in free amino acid composition in anthers of Taigu genic male sterile plants and their normal fertile counterparts were surprisingly similar to the results reported by earlier workers, using cytoplasmic male sterile materials. These results lead us to believe that the difference in the composition of amino acids is but one of the accompanying phenomena resulting from mitabolic disorders, rather than the primary reaction of the expression of the dominant gene *Ta*1.

In pistils of both types of plants, genic male sterile and fertile, no free proline could be detected at the stage when the microspores were releasing. At this stage, however, the content of glutamic acid in the pistils of the sterile plants was noticeably higher than that of the fertile plants.

References

[1] 邓景扬、高忠丽，1982，小麦显性雄性不育基因的发现和鉴定及其在遗传学和育种学上的价值。中国科学，B辑，（1）：47－57。

[2] 广西师范学院生物系，1977，水稻三系及杂交水稻花药中脯氨酸的测定。植物学报，19（1）：25－27。

[3] Ahokas, H., 1978. Cytoplasmic male sterility II. Physiology and anther cytology of *msmsl. Hereditas Lund.,* **89**: 7-21.

[4] Barnent, N. M., A. W. Naylor, 1966. Amino acid and protein metabolism in bermuda grass during water stress. *Plant Physiol.,* **41**: 1222-1230.

[5] Britikov, E. A. *et al.* 1964. Proline in the reproductive system of plants. *In* Linskens, H. F. (ed.) Pollen Physiology and Fertilization. 77-85 amsterdam, North Holland Pub. Co.

[6] Brooks, M. Z., 1962. Comparative analysis of some of the free amino acids in anther of fertile and genetic cytoplasmic male sterile sorghum. *Genetics.* **47**: 1629-1638.

[7] Dashek, W. V., S. S. Erickson., 1981. Isolation, assay, biosynthesis, metabolism, uptake and translocation, and function of proline in plant cells and tissues. *Bot. Rev.,* **47** (3): 349-385.

[8] Duvick, N. D., 1965. Cytoplasmic pollen male sterility in corn. *Adv. Genet.,* **13**: 1-56.

[9] Fujushita, U., 1964. On the free amino acids with reference to pollen degeneration in male sterile vegetable crops. *F. Fap. Soc. Hort. Sci.,* **33** (2): 133-139.

[10] Fukasawa, H., 1954. On the free amino acids in anthers of male-sterile wheat and maize. *J. F. Four. Genet.,* **29** (4): 135-137.

[11] Kern, J. J., R. E., Atkins 1972. Free amino acids content of the anthers of male-sterile and fertile lines of grain *Sorghum bicolor* (L.) Moench. *Crop Science,* **12**: 835-838.

[12] Khoo, U., H.T. Stinson, 1957. Free amino acids differences between cytoplasmic male sterile and normal fertile anthers. *Proc. Nat. Acad. Sci. B. USA,* **43**: 603-607.

[13] Linskens, H. F., J. Schrauwen, 1969. The release of free amino acids from germinating pollen. *Acta Bot. Neerl.,* **18**: 605-614.

[14] Linskens, H. F., A.M. Schrauwen, R. Koninos, N.H. Koninos, 1970. Cell-free protein synthesis with polysomes from germinating petunia pollen grain. *Planta,* (Berl.) **90**: 153-162.

[15] Ozaki, K., K. Tai, 1961. Nitrogen metabolism of paddy rice I. Free amino proline in the pollens. *Soil and Plant,* **6**: 184-185.

[16] Palfi, G. *et al.* 1981. The proline content and fertility of the pollen inbred maize lines. *Acta Bot. Acad. Sci. Hungaricae,* **27** (12): 179-187.

[17] Rai, R. K., N. C. Stokopf, 1974. Amino acid comparisons in male sterile wheat derived from *Triticum timopheeri* Zhuk cytoplasm and fertile counterpart. *Thoer. appl. Genet* (Berl.) **44**: 124-127.

[18] Sarvella, P. *et al.* 1967. Amino acids at different growth stages in normal, male-sterile and restored maize. *Z. Pflanzeng.* **57**: 361-370.
[19] Typy, J. 1963. *In* Stanley, R. G. Linskens H. F., (eds) Pollen Biology, Biochemistry and Management. p. 185. Springer-Verlag. Berlin. Heidelberg. New York. 1974.
[20] Tupy, J. 1964. *In* Linskens, H.F. (ed.) Pollen Physiology and Fertilization. pp. 86-94. Amsterdam, North Holland.
[21] Zhang Hong-qi, A. F. Croes, H. F. Linskens, 1982. Protein synthesis in germinating pollen of petunia. Role of proline. *Planta,* (Berl.) **154**: 199-203.

THE SOURCE AND UTILIZATION OF FREE PROLINE IN FERTILE ANTHERS AND THEIR RELATION TO THE ABORTION OF STERILE ANTHERS

Zhu Guanglian   Sun Chao   T.H. Tsao

(*Biology Department, Peking University*)

It is known that pollen of many plants contain large amount of free proline[7,8,13]. In some cases, free proline content could be as high as to constitute 2% of the total dry material of the pollen. Many authors consider the high content of proline to be closely related to the fertility of pollen[1,12,15,16] and to play an important role in the process of fertilization[8,9,10,17,19]. The work of our group done in the previous years with Taigu wheat on free proline in both fertile and sterile anthers points to the same conclusion. However, little is known about the source of free proline in fertile anthers. Britikov et al. (1964), based on their work with *Amaryllis hybrida*, brought up the idea that the free proline in anthers was transported from the vegetative parts of the plants, rather than being synthesized by the anthers themselves[8]. Recently, Zhang (1983), using *Petunia hybrida* as experimental material, expressed the idea that free proline in anthers had two different sources: partly being transported from the vegetative parts, and partly being synthesized by the anthers from glutamic acid[18]. These are the only papers available to our knowledge, and their conclusions do not seem to concur with each other. Furthermore, previous work on the physiological function of free proline in pollen seemed to deal mainly with the stage from maturation of microspores to fertilization, little work is known about the physiological role of free proline in the formation and development of microspores.

The present work is a continuation of the biochemical studies of genic male sterility[3]. Isotopic tracer technique is adopted to follow the source of free proline in fertile anthers, to determine the cause of deficiency of free proline sterile anthers, and to postulate the possible role of free proline in the formation and development of microspores.

Material and Methods

Plant material

After the release of tetrads, the two types of fertile and sterile plants were distinguishable by microscopic examination. When the sterility or fertility was known for the spikes of the main shoots, their corresponding tiller spikes were used for feeding experiments. At that time, the microspore mother cells within the anthers of the basal florets of the middle spikelets were just at the stage of tetrads.

The feeding period lasted from 6 to 48 hours. After these time periods, the anthers had arrived at microspore releasing stage and mononucleate stage respectively. This cytologically rather heterogeneous developmental stage is referred as phase I throughout this paper. In addition, samples were also taken to include young spikes when pollen of the basal florets at the middle parts of the spikes were at binucleate stage. This stage is correspondingly referred as phase II in this paper. After feeding, anthers of the first and second florets of the middle spikelets (amounting to 1/2 of the total spikelets) were dissected for analysis. Previous work had shown that at phase I, fertile anthers began to accumulate free proline rapidly[3] whereas the corresponding sterile anthers entered their peak of abortion, and they turned white, and ceased to enlarge[4,6].

Feeding of radioactive proline and incorporation of radioactive glutamic acid

The spikes of the above-mentioned phase I and II were cut below the nodes where the flag leaves were attached, part of the flag leaves were removed to prevent excessive water evaporation. The excised spikes with flag leaves were inserted into Knop solution containing 10 μci/ml ^{3}H-proline, and allowed to stay at room temperature (18–25°C) under continuous incandescent light of 3500 lux. New solution was added from time to time. The dosage of radioactivity was 3–10μci, varying with time of feeding. After a given period of time, anthers of the first and second florets of the middle spikelets, as well as the pistils, were dissected and subjected to amino acid analysis. There were 3 replicates for each treatment.

The vacuum infiltration experiment was performed in the following way. Anthers at both phases were infiltrated under reduced pressure with ^{3}H-glutamic acid to examine their ability to convert glutamic acid into proline. Each sample comprised 150 anthers, which were dipped directly into 2 ml 10% sucrose solution containing 25 μci ^{3}H glutamic acid. Pressure was reduced to below 50 mm/cm^2 Hg. This pressure was maintained for 5 minutes and then restored to atmospheric pressure. Repeat this process for 3 times. The infiltrated anthers were transferred to a Buchner funnel connected to suction and washed thoroughly with 10% sucrose. The anthers were finally transferred to a piece of moistened filter paper inside a petri dish and allowed to react under natural daylight at 23°C for 2.5 hours, and then dipped into 1 ml 80% ethyl alcohol for fixation. There were 3-5 replicates for each treatment.

Analysis of amino acids

1. Extraction of free proline

The anther sample was dipped into 1 ml 80% ethanol, and homogenized with a small glass homogenizer. After centrifuging at 15,000 × g for 10 min, the supernatant was collected. The extraction was repeated for four times and the extracts were combined. After removal of ethanol by evaporation at 85°C, the extract was made to a volume of 1 ml. Centrifuge for another 15 minutes, the supernatant was then ready for chromatographic analysis.

2. Chromatographic analysis of amino acid

Different amino acids were separated by means of polyamide thin layer (Nylon 66) chromatography. Before chromatography, both the sample and the standard amino acids were allowed to react with the fluorecent reagent 5-dimethylaminophthalene-l-sulphonyl chloride (abbreviated as DNS-Cl) to form DNS-amino acids[5]. Chromatography was run in one direction, the solvent system was benzene:glacial acetic acid 9:1 (v/v). Each spot contained 2 μl.

3. Determination of radioactive amino acids

Under ultraviolet fluorescent lamp, the chromatogram of the sample was compared with that of standard amino acids. The spots of proline and of glutamic acid were recognized by their respective positions. Ten spots, corresponding to 30 anthers, were then cut off separately from the chromagram, cut into small pieces and put into the counting flasks of the scintillation counter. Preparatory experiments showed that DNS-amino acids did not show quenching, and had higher counts than before derivatized by DNS.

4. Analysis of alcohol insoluble radioactive compounds

After extraction with EtOH as described above, the residue left was used for this experiment. The residue was extracted with 80% EtOH containing 3% sulponyl salicylate 6 times until the radioactive counts of the extract came near to the background counts. The residue thus treated was hydrolyzed with $6N$ HCl for 24 hr at 110°C and then counted for radioactivity. This represents the part of proline which had been incorporated into large molecules.

5. Reagents

L-(u-^{3}H)-proline, 178 mCi/mmol, L-(u-^{3}H)-glutamic acid, 32 Ci/mmol, both were products of Atomic Energy Research Institute, Academia Sinica. DNS-Cl was the product of Sigma Company. Polyamide film was produced by Yellow Stone Chemical Reagent Co. of Zhejiang. PPO and POPOP were products of Chemical Reagent Co. of Institute of Biophysics, Academia Sinica.

The scintillation counter used was Tri-Carb 460 CD, produced by Packard Co.

Results

Transport of free proline from the vegetative parts to the anthers

There are two possible sources of free proline in anthers: one is being transported from the vegetative parts of the plants; another is being synthesized by the anthers from the precursor.

The feeding experiment with ^{3}H-proline to excised spikes (with flag leaves attached) showed that during a period of 48 hours, proline entered from the vegetative parts to the anthers continuously (Table 1). It can also be seen that the increment of proline in anthers was not a constant, but was accelerated with time. This suggests that proline was not transported passively from the stem directly. It seems that proline was first taken in by other parts of the plants (mainly the flag leaves) from which it then entered the anthers by redistribution. Data presented in Table 3 will further illustrate this point. Since administrated proline can be transported from vegetative parts to anthers, we have all reasons to believe that free proline synthesized in the vegetative parts can be easily transported to the anthers, that is to say, the above mentioned first source of proline in anthers does exist in wheat plant.

From Table 1 it can be seen that the rate of proline uptake was much higher at phase I than at phase II, the former being 3.4, 4.9 and 8.0 times of the latter at 12, 24 and 48 hours after feeding respectively. This means the accumulation of free proline mainly occurred in the period from tetrad

Table 1. The uptake of free proline in fertile anthers
(cpm/mg dry wt. of anther)[1]

time of feeding (hr)	developmental phase I	developmental phase II	phase I/phase II
6	421		
12	697	205	3.4
18	1,610		
24	2,583	524	4.9
40	4,272		
48	6,132	773	8.0

1) $C \cdot V < 0.24$, $C \cdot V = \dfrac{S}{\bar{x}}$.

to mononuclear stage, a result consistent with what was reported in our previous paper[3].

Ability of anthers to convert glutamic acid into proline

Britikov (1964) considered that anthers were incapable of converting glutamic acid into proline[8], whereas Zhang (1983) held different opinions[18]. Our results obtained from vacuum infiltration experiments (Table 2) show that in fertile and sterile anthers the percentage of conversion of glutamic acid into proline was 5.17 and 4.78 respectively. Statistical analyses show that the fiducial or confidence levels of these figures were about 80%, indicating that both fertile and sterile anthers at developmental phase I possess very weak ability of converting glutamic acid into proline. At later developmental stage, phase II, under the same experimental conditions, the percentage of conversion was 16.9, indicating that at this stage fertile anthers have significant ability of converting glutamic acid into proline. When the mean rates of conversion of fertile anthers at phase I and phase II are subjected to statistical analysis, it is found that t value is equal to 11.53, much larger than $t_{0.01}(4.604)$, Therefore, the proline synthesizing ability of fertile anthers at phase II is much higher than at phase I.

Role of flag leaf in the accumulation of free proline in anthers

As shown in part I of this paper, free proline was not carried to the anthers passively through transpiration stream. To elucidate the mechanism involved in the transport of proline to the anthers, experiments were designed

Table 2. Ability of anthers to convert glutamic acid into proline at different developmental stages
(cpm/mg dry wt. of anther)

developmental stages	fertility	glutamic acid	proline	$\frac{glu}{glu+pro} \times 100$	mean rate of conversion	standard deviation
phase I	fertile	1,659	86	4.93	5.17	0.90
		1,674	110	6.18		
		1,599	74	4.42		
	sterile	2,778	180	6.08	4.78	1.59
		2,037	63	3.00		
		2,679	147	5.26		
phase II	fertile	842	175	17.2	16.9	1.52
		1,479	267	15.3		
		984	221	18.3		

to examine the role of flag leaf. The experimental material was divided into two parts. The stem segments were cut above and below the nodes bearing the flag leaf sheaths. Those cut from below the nodes had the flag leaves attached (treatment A), and those cut above the nodes were detached from the flag leaves (treatment B). In the latter case, the flag leaves were not removed, in order to keep the experimental conditions unchanged and to keep the young spikes well protected within the flag leaf sheaths. Results (Table 3) show that the accumulation of free proline in treatment A was much higher than in treatment B. At the end of 48-hour-period, the reading for treatment A was 6,132 cpm/mg dry wt[1], while that of treatment B was only 510 cpm/mg dry wt. corresponding only to 1/12 of treatment A.

Another experiment was performed to study the effect of ATP uncoupling agent, 2,4-dinitrophenol (DNP), on free proline accumulation of the anthers, using a spring variety of wheat containing $Ta1$ gene by transference. 10^{-4} mol/L DNP was added to the feeding solution, into which samples with flag leaves attached were dipped. After a feeding period of 24 hrs, the radioactivity counts in the control (without DNP added) were 1,015 cpm/mg dry wt. Whereas those with DNP added were only 458 cpm/mg dry wt., much lower than the control, and somewhat comparable to the treatment with flag leaves detached in the winter wheat experiment (Table 3). These results indicate that flag leaves are necessary for the anthers to accumulate free proline normally, and DNP is a powerful inhibitor for free proline accumulation. The removal of flag leaves had a similar effect as addition of DNP in suppressing free proline accumulation in anthers.

Situation of free proline accumulation in sterile anthers

Our previous work showed that sterile anthers of Taigu wheat lack the accumulation of free proline[3]. In order to elucidate the mechanism for this lack, feeding experiments with ³H-proline were performed with both Taigu winter wheat and spring wheat containing $Ta1$ gene. The anthers used were at microspore releasing stage. The radioactivity of the corresponding pistils was also assayed for comparison. The results were presented in Table 4.

It can be seen in both spring and winter varieties of wheat that in normal fertile plants, the anthers accumulated four times as much free proline as the pistils, whereas in sterile plants, the anthers only accumulated 60% as much as the pistils. As can be seen from the data presented, the transport

---

1) It means anther's dry weight.

Table 3. Effect of presence or absence of flag leaf on the uptake of free proline in anthers (cpm/mg dry wt. of anther)[1]

hours after feeding	treatment A (with flag leaves attached)	treatment B (with flag leaves detached)
12	697	325
48	6,132	510

1) $C \cdot V < 0.21$, $C \cdot V = \dfrac{S}{\bar{x}}$

Table 4. Comparison of radioactive free proline uptake in anthers and pistils of sterile and fertile wheat plants of spring and winter varieties (cpm/ 20 flowers)[1]

plant material	sterile plants			fertile plants		
	anther	pistil	anther/pistil	anther	pistil	anther/pistil
winter wheat	431	623	0.69	1,933	401	4.8
spring wheat	145	226	0.64	1,218	285	4.3

1) $C \cdot V < 0.19$, $C \cdot V = \dfrac{S}{\bar{x}}$

of free proline to the pistils of sterile plants was essentially normal. It would be logical to think that the lack of free proline in sterile anthers is due to some physiological disturbance in the process of transport of proline to the anthers, and the barrier is probably at the site of filaments or the anthers.

Function of free proline in microspore formation

In order to follow the fate of free proline in microspore formation and early development of male gametophyte, radioactivity assays were made on the insoluble fraction of the anthers, after being fed with ^{3}H-proline for 24 hours. Data presented in Table 5 show that in fertile anthers a considerable amount of radioactivity was found in the alcohol insoluble fraction, indicating that in early stage of microspore formation and male gametophyte development, free proline is utilized in the synthesis of protein or other physiologically active substances with large molecular weights. Whereas in sterile anthers, radioactivity in this fraction is low, amounting to only 1/6 that in fertile anthers.

Table 5. Utilization of free proline by normal fertile and sterile anthers at phase I (cpm/mg dry wt. of anther)[1]

plant material	fertile anthers	sterile anthers	fertile/sterile
alcohol insoluble	263	44	5.98
alcohol soluble	1,855	384	4.83

1) $C \cdot V < 0.18$, $C \cdot V = \dfrac{S}{\bar{x}}$.

Discussion

Our work designed to elucidate the source of proline in fertile anthers indicates that at developmental phase I and II, free proline absorbed into the vegetative parts of the plant could be transported into the anthers. This is consistent with the results obtained by Britikov and Zhang[8,18].

Our experiments with vacuum infiltration method show that at phase I, both fertile and sterile anthers possess very weak ability of converting glutamic acid into proline. At early binucleate stage (phase II), fertile anthers have considerably higher ability of conversion. This increase in converting ability may be due to the fact that at the latter stage, the anther wall tissues turned from yellowish color to green color. According to Noguchi, chloroplasts are able to convert glutamic acid to proline[14]. It is therefore possible that the conversion of glutamic acid into proline at early binucleate stage occurred in the green tissues of the anther wall, instead of within the microspores.

This part of our work concerning the proline synthetic activity of anthers is consistent with the results of Zhang[18].

The feeding experiments also show that at developmental phase I, considerable amount of proline was incorporated into large molecules in the anthers. This means that proline can serve as the raw material for the synthesis of proteins and other substances with large molecular weights. Furthermore, in the chromatogram of samples fed with ^{3}H-proline, radioactivity was also detected in the glutamic acid spot, which could be the intermediate product of the oxidative metabolism of proline. It seems that proline furnishes part of the energy source in the development of microspores. Since free proline plays such an important role in microspore formation, its lack would naturally accelerate the abortion of microspores.

Feeding experiments with radioactive proline in sterile anthers suggest that the lack of free proline in these anthers is due to some physiological barrier on the pathway of transport. Since the transport of proline into the corresponding pistils was normal, it would seem that the barrier was within the anthers or filaments. Cytological observations revealed that vascular tissues of anther and filaments were poorly developed, and the degeneration of tapetum usually occurred before the degeneration of microspores or microspore mother cells[2]. The poor development of the vascular tissues and the premature degeneration and collapse of the tapetum are likely to be the barriers which limit the entry of proline into the sterile anthers. Here we see that cytological observations coincide well with physiological analyses.

Although the above analyses indicate that free proline plays an important physiological role in the development of microspores, and its lack would affect the normal development of the microspores, we do not believe that the lack of free proline in the anthers is the primary cause of anther abortion. This idea has been expressed in a previous paper of ours[3]. It seems, then, that the expression of gene $Ta1$, via some biochemical processes, resulted in the poor development of the vascular tissues of the anthers and filaments as well as the premature degeneration and collapse of the tapetum, which make the transport of proline and other nutritive material or physiologically active substances difficult. That in turn makes the microspore mother cells or microspores degenerate,exhibiting all kinds of abnormalities during mitosis, and finally leading to the complete collapse of the microspores. In the whole process, the lack of free proline in sterile anthers is not directly controlled by *Ta*1 gene, but rather results from physiological barriers cause by it. But since proline is of vital importance for the normal development of microspores, its deficiency would naturally accelerate the degeneration of pollen.

Summary

Feeding experiments with ^{3}H-proline and infiltration experiments with ^{3}H-glutamic acid into anthers were carried out to study the sources of proline in fertile anthers, the ability of anthers to convert glutamic acid into proline, the physiological role of proline in the development of microspores and the possible causes of abortion in sterile anthers.

The following results were obtained:
1. In fertile anthers, the free proline accumulated mainly comes from the vegetative parts of the plants; however, the anthers themselves also possess some ability of converting glutamic acid into proline, especially at binucleate stage when the anther walls have turned green. This synthetic activity may be attributed to the chlorophyll containing tissues.The transport of proline from the vegetative part to anthers is an active process. It is inhibited by 2,4-dinitrophenol, and it is closely connected with the function of flag leaves.

2. In sterile anthers transport of proline from the vegetative parts was greatly hindered, and this may be the cause of proline deficiency in sterile anthers.

3. Entry of proline from the vegetative parts to the pistils of sterile plants was perfectly normal, suggesting that in the process of gene expression of *Ta*1, some physiological barriers appear in the anthers or filaments, preventing the entry of proline into the anthers.

4. In fertile anthers near microspore releasing stage, free proline administrated can serve as raw material for the synthesis of proteins or other

physiologically active substances with large molecular weights. It appears that deficiency of free proline in sterile anthers can directly affect the formation and development of the microspores, and speed up the abortion of microspores. However, we do not consider the deficiency of proline as the primary cause of pollen abortion.

References

[1] 邓景扬、高忠丽，1982，小麦显性雄性不育基因的发现、鉴定及在遗传学和育种学上的价值，中国科学，B辑（1）：47-57。

[2] 陈朱希昭,1982,太谷核不育小麦花药和花粉发育的细胞形态学观察,山西农业科学,(2)：13-16。

[3] 李祥义、邓景扬，1983，太谷核不育小麦雄性败育过程的细胞形态学研究,作物学报,9（3）：151-156。

[4] 朱广廉、曹宗巽，1984，太谷核不育小麦不育株和可育株花药内游离氨基酸成分的比较分析，植物生理学报，10（2）：185-189。

[5] 广西师院生物系,1977，水稻三系及杂交水稻花药中脯氨酸的测定,植物学报,10（2）：25-27。

[6] 陈远聪等,1975,聚酰胺薄膜层析及用于蛋白质化学分析,生物化学与生物物理进展,(2)：38-40。

[7] Bathurset, N. O., 1954. The amino acids of grass pollen. *J. Exp. Bot.* **5**: 253-256.

[8] Britikov. E.A., N. A. Musatova, S. V. Vladimirtseva, and M. A. Protsenko, 1964. Proline in The reproductive system of plants. In Pollen Physiology and Fertilization (Ed. Linskens, H. F.) pp. 78-85. Amsterdam, North Holland Pub. Co.

[9] Dashek, W. V., H. I. Harwood and W. G. Rosen, 1971. The significance of a wall-bound, hydroxyproline-containing glycopeptide in lily pollen tube elongation. In Pollen Development Physiology (Ed. J. Heslop-Harrison): pp. 194-200. Butterworths, London.

[10] Dashek, W. V. and H. I. Harwood, 1974. Proline, hydroxyproline, and lily pollen tube elongation. *Ann. Bot.* **38**: 947-959.

[11] Duvick, D. N. 1965. Cytoplasmic pollen male sterility in corn. *Adv. Genet.* **13**: 1-56.

[12] Fukasawa, H., 1954. On the free amino acids in anthers of male-sterile wheat and maize. *The Japan. J. Genet.* **29**: 135-137.

[13] Linskens, H. F. and J. Schrauwen, 1969. The release of free amino acids from germinating pollen. *Acta Bot. Neerl.* **18**: 605-614.
[14] Noguchi, M., A. Koiwai, and E. Tamaki, 1966. Studies on nitrogen metabolism in tobacco plants. VII. Δ'-pyrroline-5-carboxylate reductase from tobacco leaves. *Agr. Biol. Chem.* **30**: 452-456.
[15] Pálfi, G., L. Pintér, and Z. Pálfi, 1981. The proline content and fertility of the pollen of inbred maize line. *Acta Bot. Acad. Sci. Hung,* **27**: 179-187.
[16] Sarvella, P., B. J. Stojanovic, and C. O.Crogan, 1967. Amino acids at different growth stages in normal, male-sterile and restored maize. *Zeitschrift für Pflanzenzuchtung,* **57**: 361-370.
[17] Tupý, J., 1964. Metabolism of proline in styles and pollen tubes of *Nicotiana alata*. In Pollen Physiology and Fertilization (Ed. Linskens, H. F.) pp. 86-94, North-Holland, Amsterdam.
[18] Zhang, H. Q. 1983. Proline accumulation during anther development in *Petunia hybrida*. Ph. D. Thesis, Department of Bot. I. University of Nijmegen, Nijmegen, The Netherlands.
[19] Zhang, H. Q. and A. F. Croes, 1983. Protection of pollen germination from adverse temperatures: A possible role for proline. *Plant, Cell and Environment,* **6**: 471-476.

APPLICATION IN WHEAT CONVENTIONAL BREEDING AND RECURRENT SELECTION

Deng Jingyang Ji Fenggao
(Institute of Crop Breeding and Cultivation, Chinese Academy of Agricultural Sciences)

Wheat (*Triticum aestivum* L.) belongs to the self-pollinated crops. Its breeding method basically follows the principle of self-pollinated crops. Since the dominant male-sterile *Ta*1 gene controls the fertility of Taigu wheat and alters the way of its pollination, this gene will become an important tool for wheat breeding.

*Ta*1 gene is a dominant mutant, resulting from progenies of multiple crosses among the common wheat varieties. Therefore, in natural conditions, genes which can control the fertility of all sterile plants can only be in heterozygous state, that is *Ta*1*ta*1, which gives rise to two kinds of female gametes: *Ta*l and *Ta*l, but all the foreign pollens have *Ta*l alleles, thus, after pollination the sterile plants produce half sterile progenies having *Ta*1*ta*1 genotype and the fertile half having *ta*1*ta*1 genotype. Regardless of the cross, backcrosses or sibmating, sterility and fertility are distinct from plant to plant. There is no half sterile type and no any fertility variation caused by genetic background and climatic differences was found

The anthers of sterile plants look grey and very small and do not dehisce, belonging to non-pollen type of sterility and setting no seeds when selfing in bag. But pistils of the sterile plants are well developed and have good cross-fruitfulness with normal kernel plumpness and seed germination. No abnormality in fertility of fertile plants segregated from sterile plants is found through several generations of self-pollination[1].

So plants carrying *Ta*1 gene rely on cross-pollination to produce seeds and show the habits of cross-pollinated crops, which helps the extensive and constant gene recombinations. On the other hand, the fertile plants resulted from sterile plants can give seeds by selfing, showing characteristics of self-pollinated crops, and superior plants can be used whenever necessary in the conventional breeding Therefore, breeders can combine the advantages of both self-pollinated crops and cross-pollinated crops so as to make breakthrough in wheat breeding. Now, the main approaches of its applications are described as follows.

Application to the Conventional Breeding

In order to utilize *Ta*1 gene in the conventional breeding, it is necessary to transfer it into the parental materials having different characteristics. The way of doing this is to pollinate sterile plants with pollens of recipient varieties and make backcrosses several times. After 1 to 2 backcrosses the agronomically desirable sterile plants can be used as parents ahead of time. Then crosses are made according to different breeding methods and objectives. The principle of parent selection is the same as that of conventional breeding.

There are some advantages and reliable results when *Ta*1 gene is used in single cross, multiple cross, stepwised cross and backcross in wheat. Especially in multiple crosses, the advantages appear even more prominent. In 1959, at Davis, California, USA, when a bulk population with 8 to 10 seeds from each of 6,200 spring barley varieties collected from all over the world by USDA was used as a male parent, and another bulk population mixed of 50 male-sterile (recessive allele) materials with background genotypes of 4 varieties was utilized as a female parent, both populations were planted alternatively in rows and crossed by open and artificial pollination. After 10 to 15 generation random mating, a barley composite population having a great breeding potential was obtained and considered as a living barley germplasm pool. Apparently, for the recessive genic male-sterile materials, only when the sterile gene is homogenized through selfing can the natural cross-pollination be resumed. And it is not so easy to transfer or get rid of the sterile gene. It is much more convenient to employ a dominant male-sterile *Ta*1 gene to create a integrated population in wheat.

For an example in a integrated wheat population composed of various types of 100 materials carrying *Ta*1 gene, single crosses are counted nearly 10 thousand, double crosses are numbered about one billion, and multiple crosses can reach to astronomical figure. Within the population, in addition to the offspring involving all parents, there exist various degrees of sib-mating and half-sibbing in large amount, so the patterns of gene recombinations are extremely rich and varied[2]. The gene recombinants in the multiple cross population that a breeder can deal with in one year could be more than that he can meet in several years or even several decades in his conventional breeding work.

It is very convenient to use *Ta*1 gene to make and maintain a multiple cross population. The seeds from parents carrying *Ta*1 gene are mixed at certain rate and planted in isolated plot. Random mating between sterile plants and fertile plants takes place and then in each generation, as long as seeds are collected by given amount from individual sterile plants in multiple cross population and mixed for planting, the mixed population for next generation can be composed. Since half fertile plants can be segregated from each of generations, desirable fertile plants from any generation can be selected and treated according to the conventional breeding procedure. When making up the initial bulk population, the proportion of seeds of agronomically desirable parents and parents carrying recessive objective traits should be higher than the others.

For Recurrent Selection

The recurrent selection for self-pollinated crops like wheat should involve three parts of contents: 1) creating a initial mixed population; 2) improving the population by recurrent selection; and 3) utilizing the improved population. The key point lies in realizing gene recombinations in a specific artificial population → selecting the superior gene recombinants → gene recombinations → selection → , moving forward in cycle and making constant progress. The frequency of desirable genes in population can be gradually raised through cycle selection and the undesirable gene frequency will decrease generation by generation. Thus the genetic base in the whole population is enhanced, the desirable genes recombined are increased and the probability of integrating more desirable genes into a plant is raised, which is to create an excellent genetic foundation for breeding program.

The recurrent selection helps to break the unfavourable gene linkage. There are two considerations: Firstly, a bigger population should be maintained for selection in favour of increasing the number of crossover individuals; secondly, selection should be conducted under certain pressure for several years to accumulate crossover plants, so the "superior-superior" linkages replace the original "superior-inferior" ones.

Since the genotypes between plants which constitute the "pure line" variety are identical, the total capacity of genes remains unchanged regardless of the size of population and is equal to a capacity of gene pool of any individual plant. On the contrary, in the population composed of heterozygous plants with different genotypes, the larger the size of population, the bigger the capacity of gene pool and the higher the heterozygosity of each plant, the bigger the capacity of gene pool too. Because every sterile plant within a recurrent population relies on the cross pollination for seed setting, each offspring is highly heterozygous and their genotypes between plants are different from each other. Accordingly, one recurrent population can contain numerous genes that can affect different traits, and many genes influencing the same trait, no matter whether they are homogeneous and allelic or nonallelic, can exist together. For it is possible to incooperate different nonhomogenous gene affecting the same trait into a population. In certain sense, this kind of population is to become an artificial gene pool for this trait. These nonhomogenous genes are constantly to be recombined, consequently the "living gene pool" is extremely useful for selecting various extraordinaly type of better plants. As far as the comprehensive traits is concerned, it is also helpful to select individual plants with the better composite traits.

After accomplishment of cross plant in the conventional breeding program, the gene exchanges no longer occur among individual plants. With the increase of generation of selfing and the degree of homogenization, the capacity of gene pool become smaller and smaller, many genes are inevitably lost and as a result, the selection effect is impaired.

The way of composing a recurrent population is similar to that of making multiple crosses to form a mixed population. The essential difference between the recurrent selection and the multiple crosses are: the former is a continuous process and has certain selection pressure so that the genetic bases could be gradually enhanced, i.e. the genetic quality in whole mating population can be continuously improved, whereas the latter is nothing more than achieving gene recombination. The newly composed population is called primary initial mixed population. Then free pollination is carried out for 3 to 4 generations so as to make extensive gene recombination and enlarge the population to the required size as well. At the moment, the better gene recombinants are still too limited and no strict selection on the sterile plants is done. No pathogen innoculation is conducted to the population for disease resistance selection at present. The real recurrent selection should be started from next stage, i.e. secondary stock population.

In order to prevent the spread of unfavorable genes carried by undesirable fertile plants in the population, selections on the male gametes should be started earlier. The sterile plants with undesirable traits only have their impacts on themselves, so harvest not their seeds to eliminate the undesirable genes that they carry. However, fertile plants with undesirable traits may affect the surrounding sterile plants that receive their pollen grains, for their paternal traits can not be determined when seeds are collected from sterile plants. Therefore, it is very important to wipe out undesirable plants rigorously during the period from heading to flowering, but this point is easy to be ignored.

The different selection methods are adopted concerning the objective traits whether they are dominant or recessive, expressed before flowering or after flowering, and can be determined at present generation or not.

Blend selection (effective for the dominant traits)

The practices of corn breeding have proved that blend selection is effective to the quantitative traits and qualitative ones[3].

1. For the traits like plant height, tillering ability, seedling disease, heading stage, cold tolerance, drought resistance, salt and alkali-tolerance, which can be selected before flowering, one should not only make full use of the environmental selection pressure but also eliminate undesirable plants (regardless of sterile or fertile plants) at the key stage when the traits are most visibly expressed. So, the better genetic quality of both male and female gametes involved in flowering and pollination are ensured. After mature stage, seeds collected from desirable sterile plants are mixed and planted to form a new population in which same way of selection is carried out. In order to control the population size, about 10 kernels (depending on selection pressure) are collected from main culm of the selected sterile plants every generation and mixed for next cycle. The rest seeds of desirable sterile plants can be sown separately and the segregated desirable fertile plants may be used for pedigree selection. The desirable fertile plants selected with the population should be grown separately in lines and used for breeding program.

2. For the characteristics that can be expressed after flowering, like late disease resistance to scab, rust, smut, powdery mildew and resistance to dry and hot wind, drought resistance, rapid kernel filling, heavy ear, large kernel, quality and so on, the pollinator should be sown in another place for the sake of ensuring the genetic quality of male parents.

The sterile plants and fertile plants in recurrent population have sibship and equal averaged genetic level. Nevertheless, desirable sterile plants only have a half genetic contribution to their offspring while good fertile plants rely on selfing for setting seeds which containing whole set of genes from their parents. Accordingly, seeds from the selected fertile plants in recurrent population are good pollinator for next cycle.

From the blend population, both desirable sterile plants and fertile plants are selected at the same time Their seeds are separately blended and grown alternately in lines All fertile plants in the sterile plant-seed line are uprooted from heading to flowering. The undesirable individual plants in the fertile seed-plant must ripe out at any time.

There are three sources of pollinators in the next cycle: (A) to reselect desirable fertile plants from the fertile-plant lines (it is inadvisable to do it for many generations otherwise genetic base will become increasingly narrower); (B) to mix and sow part of the selected sterile-plant-seeds from each generation, from which desirable fertile plants are chose for next pollinator; (C) to blend and sow a fixed amount of seeds from better materials or strains which are selected for breeding.

3. Concerning the quality characteristics, like high protein content, etc; First, a primary selection according to comprehensive traits is carried out in the fields (including both sterile and fertile plants). Then seeds are threshed and labeled plant by plant. Part seeds of each plants are taken for protein content analysis. The definite amount of seeds from sterile plants with high protein content are mixed and sown as next female parent line. The seeds from corresponding fertile plants with high protein content are blended and grown for next cycle male parent. The selection can be conducted according to the way described in (2). Analysis can be done only for sterile plant seeds, and recurrent selection may follow the way (1).

The above methods are simple for practise. Although the selection response is not high, the accumulated selective advance will become considerable for selection is conducted generation after generation. Nevertheless, selection in this way is only based on the phenotype at present generation, which is subject to disturbance of environment. It has no effect on the recessive genes.

Plant line selection for every other generation (effective for the recessive traits)

In the free-pollinated secondary population, desirable sterile plants are selected, threshed and numbered plant by plant. The seeds from each plants are grown in the square-shaped plots. The distance between plots should be large enough to reduce pollen spread. In the superior plot, better sterile and fertile plants are selected and bagged for mating, and the harvested seeds are used for next cycle mixed population. In this way, the plant plot and mixed planting are carried out alternately. For this method, plant line should be selected first and then the individual plant should be selected within line, so we could get higher accuracy. The sibmating within line plot provides an opportunity for the homogeneity of recessive genes. So the possibility of losing the desirable recessive genes covered by the undesirable dominant genes may be deminished. Certainly the homozygous recessive gene can only express in the following generation, i.e. in mixed population.

Many kinds of recurrent selection method can be derived from the above two. As to the method suitable for the undeterminable traits, it has not much use at present and is not discussed here.

In the process of recurrent selection, the recurrent populations for different ecological types should be composed according to ecological environment in different regions. The recurrent selection can be seperately oriented not only to a single trait, like early maturity, disease resistance and high protein content, etc., but also to the comprehensive traits.

The mixed population obtained after several cycle selection can be considered as a improved population. Since the favourable gene frequency has been enhanced within the population, one should not hastily introduce new genes outside of population. the cross-pollination and gene recombination should be controlled within the recurrent population as fully as possible, because a lot of genes, regardless of favourable genes or undesirable ones, will be inevitably brought into the improved population by the alien gene carrier.

If a new gene has to be introduced, the gene's carrier should be crossed with desirable plants from the improved population and the resulting desirable offspring can be chosen and added to the recurrent population.

Not introducing new genes outside of the recurrent population, the effeciency of selection, for the qualitative characters controlled by a few genes at the beginning may be very high but advances very slowly later, while it is always efficient to many quantitative characters related to yield production. Constant raising the level of expression of the objective quantitative is one of the prominent advantages of recurrent selection method and can not be reached by any other means in breeding.

Among different improved population with high genetic basis, gene flow not only will not make the genetic level lower, but also is able to enrich their genetic background. Therefore, if possible, at the beginning, it is necessary to constitute more recurrent populations and then gene recombinations between the populations can be achieved according to demand of breeding program.

Each improved population in each generation can be divided into three parts by selection: superior fertile plants, good sterile plants and the remaining population, each of which can be used for different purposes. The seeds of good sterile plants can utilized in the continuous recurrent selection. The better fertile plants may be employed as the basic materials for new variety breeding, and considered as F_1 hybrids of multiple crosses. So their seeds (as F_2) set by them could be planted in mixed sowing and the plant lines could be established from F_3 generation. In case there are a number of the recurrent populations with large volume, the segregated fertile plants from each generations will be numbered by thousands so that single seed descendant method of selection may be adopted. In this way from F_2 generation, only some seeds or even one seed are harvested from individuals and planted in mixed sowing. The same way is to be continued up to F_5 or F_6, and then individual plant selection is carried out to set up plant lines.

Large number of fertile plants come out of each recurrent population and sufficient materials can provided for selection. As the generations advance forward, the genetic levels of the segregated fertile plants can be raised, so a series of new strains can be selected constantly.

With the aid of anther culture technique and parthenogenesis, the selected sterile and fertile plants can be fully used and the breeding cycle can be shortened greatly so as to speed up the progress of new variety breeding.

The anther culture breeding has been widely applied in China, but not many varieties having large growing acreage are broght with this method. One of the important reasons is from anthers themselves. If the genotypes of male gametes are not good, it is hopeless to breed a variety with better traits. The first important work for anther culture is to select the anthers that contain high proportion of pollen grains having desirable genotype. The recurrent improved population has high frequency of favorable genes and abundant genotype. The anthers from their fertile plants are desirable materials for anther culture.

After the selection of sterile and fertile plants, the rest mixed populations can be planted in mixed sowing for several generation, and then fertile plants are seperated into groups according to the main agronomical traits (such as plant height, and maturity length of awn etc.). After selfing and culling several composite varieties with diverse traits can be obtained. These composite varieties produced by biological mixture have rich genotypes as compared with those produced by mixture, so they have wide adaptability and high yield potential and may be well fit to the low yield producing regions.

Another recommended way to breed composite varieties is to select the fertile plants directly from the improved population.

Borlaug and Wiebe recommended enthousiastically using the genic male sterility in breeding. Sorrells and Fritz designed a flexible and variable scheme using a dominant male sterile gene in recurrent selection of self-pollinated crops. We believe that some more reasonable and efficient breeding programs would be worked out as the dominant male sterile $Ta1$ gene has been used extensively in wheat breeding. Furthermore, new breeding programs and breeding methods will advance the use of $Ta1$ gene more thorough and efficient in wheat breeding.

References

[1] 邓景扬、高忠丽, 1980, 小麦显性雄性不育基因的发现与利用—太谷核不育小麦鉴定总结。 作物学报, 6 (2): 85-98。
[2] 吴兆苏, 1983, 美国大麦、小麦雄性核不育的研究利用动态。麦类作物, (1): 1-6。
[3] 杜尔宾等著, 刘杰龙等译, 1976, 植物育种的轮回选择, 农业出版社。
[4] Athwal, D.S. and N.E. Borlaug, 1976, Genetic Male-sterility in Wheat Breeding. *Ind. Jour. Genetics and Plant Breeding,* 27(1): 136-142.
[5] Sorrells, M.E. and S.E. Fritz, 1982, Application of A Dominant Male-sterile Allele to Improvement of Self-pollinated Crops. *Crop Science,* 22: 1033-1035.
[6] Wiebe, G.A. 1975, The Challenge Facing Barley Breeders Today. Proc. 3 Int. Barley Gen, Symp. pp. 1-10.

STUDIES ON THE COMBINATION OF BASE POPULATION IN POPULATION IMPROVEMENT

Wang Jinxian Wu Kezhong Zhou Mingfu

(*Jilin Provincial Academy of Agricultural Sciences*)

Recurrent selection is an effective means for crop population improvement. With this method remarkable results have been achieved in such crops as maize and other cross pollinated crops in the past few years. It can be expected that because of advance in the study of recurrent selection, particularly the discovery and identification of the dominant genic sterile germplasm in *Ta*1 wheat[1], wheat population improvement should achieve remarkable results.

Based on a population with a wide range of inherited characters and high frequencies of beneficial genes, recurrent selection is used in accordance with breeding goals. After several cycles of selection and intercross, undesirable gene linkage will be broken down, increasing the probability of accumulation and recombination of desirable genes. A population with rich inheritable variations should be formed at first, and this population should be able not only to intermate and recombine, but also to self pollinate. Owing to the discovery of *Ta*1 genic sterile materials, wheat as a self pollinated crop becomes able to cross through the sterile plants and through the fertile plants[3], providing conditions for recurrent selection. Since wheat is a self pollinated crop, its varieties or strains are homogeneous populations propagated by a homozygous genotype. Therefore, any varieties cannot be directly used as base population for recurrent selection. It must be at first coincided with the requirement of a base population. Questions occur such as what kind of materials (selection of parents) and how much of the materials (number of parents) will be used and how to combine the population (forms of combination). These problems, especially concerning the populations combined by dominant *Ta*1 genic sterile gene are rarely described in existing papers. Hallauet (1981) made a comprehensive description on recurrent selection[3], and Wu zhousu(1983) reported the studies and utilizations of genic sterile genes of barley and wheat in United States[2]; neither of them, however, dealt with these problems. The combination of base population and breeding of the 'breakthrough' varieties with superior synthetic characters, will be discussed in this article.

Selection of Parents

Though the parental selection of base population may draw on the experience of conventional breeding, most cases in conventional breeding are combination breeding with only two to four parents. It combines desirable characters of male and female parents giving them to their offspring mainly by means of one cross or 2-3 successive matings. Generally, on the basis of keeping the desirable characters of the original varieties, one or two of their characters will be improved. From the viewpoint of cytogenetics, only one to three pairs of chromosomes or chromosome segments in parental gene groups could be replaced. The selection of their parents is mainly based on the principle of character complementation. The recurrent selection is essentially a transgresssion breeding, so there are more parent stocks used for it. It uses both the law of seperation, recombination, and the polygenic additive effect to create recombinators with transgressive characters. Through ramdom mating among individuals, all the parents genes are transfered to hybrid population, thus forming a heterogeneous population with a lot of stored desirable genes, which can be used as basic materials for recurrent selection. Therefore, parents for the base population of recurrent selection should be selected not only according to the principle of conventional breeding, but also according to its own performance, hence the following points must be considered:

1) The transgressive inheritance of quantitative characters is determined by polygenes which differentiate locus on chromosomes or by micro cumulative polygenes. If we want to get this accumulation, the general combining ability of objective characters must be amphasized when considering the selection of parents and their complementary characters. The additive effect of some characters and the occurrence of transgressive inheritance are determined by their general combining ability. So attention should be paid to the selection of phenotypes, as well as to understanding the difference in their genetic composition and genetic origin.

2) The main aim of recurrent selection is to raise the frequencies of desirable genes controlling characters, consideration must be given to it dur-

ing the parental selection. Parents with desired traits should be prominent since recombinations of these parents will bring about better integrated recombinations. But parents with undesirable traits with high heritability should not be used. On the other hand, we must raise the ratio of parents with desirable traits but with poor competition characteristics, such as short stalk, compact plant type, small leaves and poor tillering. If their frequency is low, the desired characters will often disappear in competition after a few generations of intercrossing.

3) There are two kinds of parents that form the base population. One is a prominent type in yield, disease resistance and superior in agronomic characters; the other is dominant in adaptation to the local physiological and ecological conditions, for example, rot resistance, high temperature resistance, rapid filling, or long function of flag leaf. These characters meet the requirement late growth period in high temperature and rainy areas. An improved variety with high yield, and strong resistance to negative factors must be based upon the adaptability to specific ecological requirements.

4) An inference can be drawn from the above disscussion that if excellent individual plants selected from F_2 of highly variable population or from excellent F_3 families of better combination by conventional breeding can be used as parents for the mixed population, combining with sterile plant of Ta1 type that possesses desirable synthetic characters the results will be even better.

Number of Parents

The number of parents used to form the population was detailed by Wang Zhenfu[2]. It was deduced that under random mating, the frequency of each parent combination appearing in each generation of population is:

$$P = \frac{n^{2m} - (n-1)^{2m}}{n^{2m}} \times 100$$

(n = parent number, m = times of crossing)

According to this formulae, the more parents used, the lower the frequency of each parent and the smaller the increasing value obtained. He held that 20–40 parents would be suitable. Rasmusson suggested that to maintain a diversified inheritance, at least 8 parents should be used. We suggest that 10–15 parents would be more suitable, because parent number is determined by the expected accumulation and complementation of objective characters. In general, not more than three or four characters of varieties available need to be improved. Three to five improved varieties or lines in combination with seven to ten breeding lines with outstanding characters will be quite enough to produce a population. The famous wheat varieties with dwarf stalk and high yield such as Gains from the U.S. and Mexican wheat were developed with only about ten parents from Nonlin 10 × Brever, after several cycles of matings and accumulation. This proves that the number of parents required need not to be very large.

Depending on the breeding requirements for improved varieties, eight-parent crossing population with 50–5000 different genotypes can be produced after intermating for three successive generations. The number of parents required is at least ten (containing 45 eight-parent crossing materials) and at most is 15 (containing 6,435 eight-parent crossing materials). When parents are less than ten, the range of genetic variation is too narrow. Eight-parents, for example, can only be developed into an eight-parent type materials after intercrossing for three generations. After that, they can only be limited in segregation and recombination of eight parent genotypes. If there are too many parents, the population needed would be too large for all possible recombination to stabilize. Identification and selection, and cross pollination will all be limited. With 20 parents and only three random intercrosses, the theoretical number of individuals in a population goes up to 25.6 billion. With 125,970 eight-parent crosses, each cross will get a sterile plant, so we can compact the population size to one twenty thousandth. There still will be 1,280,000 individuals, covering an area about 6,350 m^2. In addition, the effect of recurrent selection is mainly determined by the frequencies of superior genes; the more parents and varieties with specified traits, the lower frequencies of superior genes in a population.

From the viewpoint of additive effect and theory of transgressive inheritance, if the parents are all heterozygous materials with high general combination ability, 10–15 parents will meet the requirement of transgression of objective characters and recombination of desirable genotypes.

With complex breeding goals combining such characters as early maturity, high yield, and good quality, and resistance to diseases or pests, more parents may be required, when number required is more than 15, it is advisable to combine various general characters first, and then to find ways to add specific traits to the improved population.

Patterns of Combination

The key to the combination of base population lies in full random intercrossing among the individuals. There are only two basic patterns of combination, random intermating of mixed individuals and half-diallel or incomplete diallel crossing. They can also be divided into two forms, in terms of whether the parents used are all $Ta1$ type varieties.

Random intermating of mixed individuals

1. Taking seeds from $Ta1$ type materials tested for combination ability, mixing and planting the seeds and random intermating among individuals to form base population.

This system is also suitable for organizing population with different characters.

2. Mixture of male parents by using nontransfered materials as male parents to organize the population.

Take an improved variety (or the mixture of 2–3 varieties) of $Ta1$ type wheat as female parents, and mix them with seeds from male parents with specified features tested for combination ability; plant them alternatively with female parent rows; harvest the sterile plants from the female parents rows; mix the seeds equally; intermate the mixed individuals.

Half-diallel or incomplete diallel crossing

1. Half-diallel crossing of $Ta1$ type materials. This system also uses parents tested for combination ability. At first, a half diallel crossing is used. Then conduct a random intermating of mixed individuals. Those populations in case with only a few parents can have two more half-diallel crossings and then a random intermating. In order to reduce the number of combinations, we select the combination of four parents at the second cycle of mating based on the first cycle of half-diallel crossing. If there are ten parents, 45 single cross combinations are produced at the first cycle of half-diallel crosses and a second cycle could produce cycle 990 combinations. If we select only the four parent combinations, only 210 combinations will be required.

2. NCII mating takes the seeds of transfered varieties of $Ta1$ type material as female parents, and crosses with materials with outstanding traits tested for combination ability (offsprings from one female parent with several male parents). Seeds of sterile plants in female parents rows are harvested, mixed and intermated with the individuals of the next generation.

The methods mentioned above can be used according to the conditions. Here some problems on combination will be discussed as follows.

Discussion

1. Half-diallel is recommended in the first combination of base population instead of intercrossing of mixed individuals.

The ideal organized populations are those which possesses the lowest frequency of similar inbreeding type and the highest recombination abilities. The mixed individuals, due to the parents taking part in the intercrossing from generation to generation, may obtain a great amount of AAAA, AABB AAAK and AABK after conducting double random intercrosses, perhaps there are many inbreeding types or the similar ones which we needn't. At the second cycle of intercrossing, more $n^2(2n-1)$ inbreeding individuals and similar types are produced than in complete diallel crossing without parent participating (n is for number of parents). For example, if eight parents intercross twice in successive generation, there will be 960 individuals of the same type, accounting for 23.4% of the total population. This is useless for expanding the population size.

A population is formed by eight parents, which are used to conduct double random intercrosses of mixed individuals, has 4,096 individuals. When half-diallel is adopted at first and followed by the random intercross, there are 784 individuals in the population, or only 19.1% as many. If half-diallel is continually adopted twice, the population has 378 individuals, or only 9.2% (see Table 1).

If eight parents have made double random mating, only 70 four parent combination types can be obtained, and of the 4,096 individuals, 1,681 are individuals of four parent combination, accounting for 41%. If a half-diallel cross is followed by a random intermating, of the 784 individuals, there are 420 individuals of four parent combination, accounting for 53.6%; if half-diallel crosses twice, of the 378 individuals, there are 210 individuals of four parent combination, accounting for 55.6%. (Table 1).

Therefore, we suggest that first combinations of base populations would be better done with half-diallel cross. In this way, we can decrease the parent types and similar inbreeding types from progenies, increase the probability of multiparent combinations, compact the population size, and generally improve the combinations and retain desirable parent genes.

No matter what method is used, in the stage of mixed individual intercross, the gene frequencies of progeny and methods adopted to specific conditions must be considered. All programs cannot follow the same pattern in different conditions.

Table 1. Comparison of the ratio of population size with
the highest parent combination type of each
mating in four mating forms

No. of parent	mating forms[1]			1st		2nd				3rd				
	1st	2nd	3rd	No. of popul. indiv.	No. of single comb.	population		4-parent cross		population			8-parent cross	
						No. of indiv.	%	type	No. of indiv. %	No. of indiv.	%	type	No. of indiv.	%
8	I	I	I	64	28	4096	100	70	1680 41.0	16777216	100	1	40320	0.24
	h	I	I	28	28	784	19.1	70	420 53.6	614656	3.66		2520	0.40
	h	h	I	28	28	378	9.2	70	210 55.6	142884	0.85		630	0.44
	h	S	I	28	28	70	1.7	70	70 100	4900	0.03		70	1.4
10	I	I	I	100	45	10000	100	210	5040 50.4	100000000	100	45	1814400	1.81
	h	I	I	45	45	2025	20.3	210	1260 62.2	4100625	4.1		113400	2.77
	h	h	I	45	45	990	9.9	210	630 63.6	980100	1.0		28350	2.89
	h	S	I	45	45	210	2.1	210	210 100	44100	0.04		3150	7.1
12	I	I	I	144	66	20736	100	495	11880 57.3	429,981,696	100	495	19958400	4.6
	h	I	I	66	66	4356	21	495	2970 68.2	18,974,736	4.4		1247400	6.6
	h	h	I	66	66	2145	10.3	495	1485 69.2	4601025	1.07		311850	6.8
	h	s	I	66	66	495	2.4	495	495 100	245025	0.056		34650	14.1
13	I	I	I	169	78	28561	100	715	17160 60.1	815730721	100	1287	51891840	6.4
	h	I	I	78	78	6084	21.3	715	4290 70.5	37015056	4.54		3243240	8.8
	h	h	I	78	78	3003	10.5	715	2145 71.4	9018009	1.11		810810	9.0
	h	S	I	78	78	715	2.5	715	715 100	511225	0.06		90090	17.16
14	I	I	I	196	91	38416	100	1001	24024 62.5	1475789056	100	3003	121080960	8.2
	h	I	I	91	91	8281	21.6	1001	6006 72.5	68574961	4.65		7567560	11.0
	h	h	I	91	91	4095	10.7	1001	3003 73.3	16769025	1.14		1891890	11.3
	h	S	I	91	91	1001	2.6	1001	1001 100	1002001	0.067		210210	21.0
15	I	I	I	225	105	50625	100	1365	32760 64.7	2562890625	100	6435	259459200	10.1
	h	I	I	105	105	11025	21.8	1365	8190 74.3	121550625	4.7		16216200	13.3
	h	h	I	105	105	5460	10.8	1365	5460 75.0	29,811,600	1.16		4054050	13.6
	h	S	I	105	105	1365	2.7	1365	1365 100	1863225	0.072		450450	24.2

1) Mating forms, I-I-I refers to three times of open pollinated intercrossing; h-I-I system refers to the first time is half-diallel, the second and the third times are intercrossing; h-h-I. system refers to the 1st and the 2nd times are continual half-diallel, the 3rd is intercrossing; h-S-I system refers to 4-parent combination type at the 2nd time.

Table 2. Compacting the population size by compact coefficient in the third intercrossing stage of four kinds of mating forms

No. of parent	mating forms			3rd intercrossing			compacting coefficient	compacted			remark
	1st	2nd	3rd	No. of pupulation & indiv	8-parent cross type	No. of indiv.		population size	%	No. of 8-parent indiv.	
8	I	I	I	16777216	1	40320	a 1/10000	1677	100	4	
	h	I	I	614656		2520	b 1/630	975	58.1	4	the compacted
	h	h	I	142884		630	c 1/157.5	907	54.0	4	population
	h	S	I	4900		70	d 1/17.5	280	16.7	4	size is
10	I	I	I	100000000	45	1,814,400	a	10000	100	180	equal to
	h	I	I	4100625		113400	b	6509	65.8	180	theoretical
	h	h	I	980100		28350	c	6223	62.2	180	planting
	h	S	I	44100		3150	d	2520	25.2	180	coefficiency
12	I	I	I	429981696	495	19958400	a	42998	100	1980	
	h	I	I	18974736		1247400	b	30119	70.0	1980	
	h	h	I	4601025		311850	c	29213	67.9	1980	
	h	S	I	245025		34650	d	14000	32.5	1980	
13	I	I	I	815730721	1287	51891840	a	81573	100	5148	
	h	I	I	37015056		3243240	b	58754	72.0	5148	
	h	h	I	9018009		810810	c	57257	70.0	5148	
	h	S	I	511225		90090	d	29212	35.8	5148	
14	I	I	I	1475789056	3003	121080960	a	147578	100	12012	
	h	I	I	68574961		7567560	b	108849	73.7	12012	
	h	h	I	16769025		1891890	c	106470	72.1	12012	
	h	S	I	1002001		210210	d	57257	38.6	12012	
15	I	I	I	2562890625	6435	259459200	a	256289	100	25740	
	h	I	I	121550625		16216200	b	192938	75.3	25740	
	h	h	I	29811600		4054050	c	189280	73.8	25740	
	h	S	I	1,863,225		450,450	d	106,470	41.5	25740	

2. In organization of population, this method should ensure that many eight parent combinations will be obtained after three cycles of intercrosses, and recurrent selection can begin.

Recurrent selection should not be initiated too early since there are only two parent combinations at first intercrossing and four parents at second intercrossing so the accumulating characters and recombination are rather few. If we initiate a recurrent selection at this time, it is difficult to select desireable materials and to lose desirable genes or even lose the value of the whole population.

3. Size of population and its controlling parameter after three random intercrossings of 10-15 parents, the smallest population contains one hundred million individuals and the largest contains more than 2.56 billion individuals. It is impossible to deal with such a large population. Thus it is neccessary to make the population more compact. How shall we reduce the population so that we neither lose good genes nor breakdown the population genetic equilibrium?

(1) To compact the population size in accordance with the theoretical coefficient. Based on probability calculation, when intermating for one to three times, the size of population, the highest number of parent combination and its individuals at each time of mating are given in Table 1. We can see that if eight parents are randomly intercrossed three times, the total number of individuals in the population is 16,777,216, but only one kind is an eight parent combination type. These number 40,320, that is to say, after three random intercrossing, the number of individuals of eight parent combination is 40,320 times the number of eight parent types. This provides a basis for compacting the population size by eight parent of random intercrossing. If each eight parent combination type gives one sterile plant only, we can compact the population with this probability: the total individual number of population should be divided by 20,160 (as half the fertile plants involved). In order to prevent losing any combination type, two sterile plants should be used as one. The compact coefficient of the population should be 20,160/2 = 10,000, that is to take 1/10,000 as the control coefficient of population. Being compacted by this coefficient and three random intercrossings, the population formed by 10 parents have 10,000 plants, by 12 parents have 42 998 plants, by 15 parents, 256,289 plants. (see Table 2) Considering genetic complication of the combination of multiparents, we should keep 10-20 individuals of sterile plant for each type. The compact coefficient will become as 20,160/10 - 20,160/20, that is one two-thousandth or one thousandth.

With this same principle, the half-diallel cross at first cycle is followed by two random intermatings. Each eight parent combination type has 2,520 individuals and their limit compact coefficient is 2,520/4 = 630. The population formed by 10 parents has 6,509 individuals, 12 parents give 30,119 individuals, and 15 parents give 192,938 individuals. For the population with the first and second cycles of half-diallel cross and the third cycle of a random intercrossing, the limit compact coefficient of the population is 630/4 = 157.5. The population formed by 10 parents, after compacting, has

6,223 individuals, 12 parents give 29,213 individuals, and 15 parents give 189,280 individuals (see Table 2).

(2) With the reduction at the second cycle according to four parent type of combination.

We can see from Table 1 that, of the three forms of combination mentioned above, the form with two half-diallel crosses at the first and second cycles followed by random intermating at the third cycle has the smallest size. However if the parents used are more than 11, the manual work of artificial crossing is too much to carry out. For instance, at the period of second half-diallel crossing, 11 parents need to make 1485 crosses, 12 parents need to make 2,145 crosses, and 15 parents need to make 5,460 crosses. We have designed another combination system that uses only the different four parent types at the time of second combination. So only 495 crosses will be needed when 12 parents are used, 715 crosses for 13 parents, 1,001 crosses for 14 parents, and 1,365 crosses for 15 parents (reference to Table 1). The population formed by this method is only 21–25% of the population formed by the previous method. At the third cycle of combination, the size of the population from 10–15 parents is reduced to 4.5–6.2%. While combining the population by this method the third time, a compact coefficient is still needed which is 70/4 = 17.5, namely 1/17.5. The compacted population that formed by 15 parents can get 106,470 individuals, and the double half-diallels after being compacted still get 189,280 individuals.

Evaluation of four systems of organized population

After compacting the population of the third intercrossing by the theoretical coefficient, the population sizes produced by the four combination systems are given in Table 2. The figures in the table show that in the case of 15 parents, the largest population size of I-I-I (I is for intercrossing) has only 0.26 million individuals; a 1,000 m^2 seedling nursery is quite enough for that. So, it is rather easy to conduct and the same for the other three systems. But considering labour needed and combining effect, each of these four systems has its good points and should be adopted according to the conditions.

1. I-I-I system

I-I-I system is very easy to carry out and suitable for the organized population with parents less than 12. However, attention must be paid to the anthesis and the plant height for if their differences are too large, there will be problems in incomplete combination.

2. h-S-I system

h-S-I system (h is half-diallel and S is selection); its compacting effect is outstanding, suitable for populations from more than 14 parents.

3. h-I-I system and h-h-I system.

h-I-I system and h-h-I system suitable for organized populations with less than 14 parents; both have similar size after compacting. For example, when parent number is eight, the difference between them is about 7%; when the number of parents increase to 15, their difference is less than 2%. But the h-h-I system is difficult to carry out, because the manual work needed at the second half-diallel cross is too much. Considering the difficulty, it is better to use the h-I-I system.

The items mentioned above were theoretical deduction based on progeny segregation and recombination law of conventional breeding. There is need for experimental data and practice. Any response from our readers concerning questions raised or possible errors will be welcomed.

References

[1] 邓景扬等,1980,小麦雄性显性不育基因的发现和利用,作物学报 6 (2): 85-98。
[2] 吴兆苏,1983,美国大麦、小麦雄性核不育的研究利用动态,麦类作物,1: 1-6。
[3] 王进先等,1983,太谷核不育小麦在育种中应用方法的探讨,吉林农业科学,3: 23-29。
[4] Hallauer A.R., 1981, In plant Breeding II, P. 356. Iowa State. Univ Press.
[5] 王振富等,1981,显性不育基因在小麦轮回选择中应用,山西小麦通讯,(3): 25-35。

SELECTION AND MATING IN RECURRENT SELECTION

Zhang Shaonan
(*Institute of Food Crop Research, Agricultural Academy of Shanxi Province*)
Ji Fenggao
(*Institute of Crop Breeding and Cultivation, Chinese Academy of Agricultural Sciences*)

In this paper we make some suggestions on the application of the dominant male sterile gene to the recurrent selection and mating in wheat.

The sterile plants can only be produced from the sterile plants by cross fertilization, thus all sterile plants belong to F_1 generation. Among the recurrent population, with random matings and a large number of mutiple alleles, better selection response could not be obtained from the sterile plants. The artificial sib-mating between plant rows is not so good as the self-pollination of fertile plants in the speed of homogenization. In addition, external factors affecting pollination will influence the fertility of sterile plant and will affect seed plumptness, kernel weight and the appearance of plant maturity. These factors will impede the selection of sterile plants. Therefore, the key object of selection is not the female gametes provided by the sterile plants, but the male gametes of fertile plants. Concerning the different sources of male gametes and selection methods, the following four recurrent selection programs can be formed (Fig.1):

Program I: This is a mass selection, in which plot I and plot IV are utilized. In each cycle, the sterile plants with better performance are chosen from plot I and the harvested seeds are blended for the next cycle which continues the random mating and population selection. The best fertile plants from each cycle can be used in plot IV as breeding materials for new varieties.

Program II: Plot I provides seeds for the maternal lines and all fertile plants should be eliminated before flowering. Plot II provides seeds for the paternal lines. The breeding materials are also chosen from the fertile plants

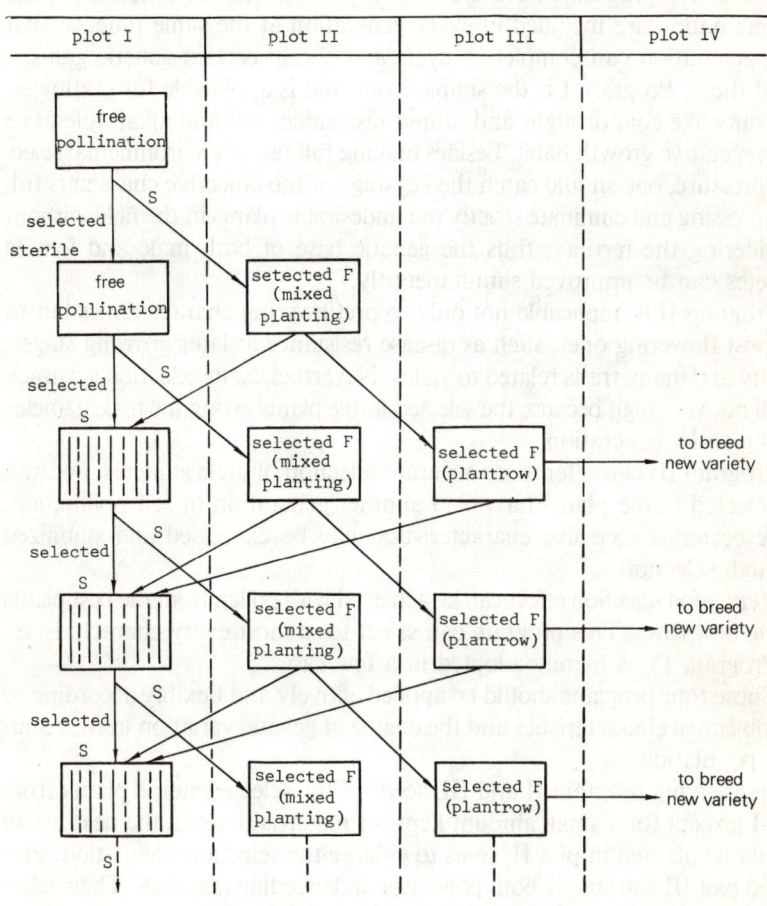

S: sterile plants, F: fertile plants
—: row of father, ······ : row of mother

in plot II.

Program III: Seeds come from plot III and the rest follows Program II.

Program IV: The materials from plot IV and other advanced generations, the best fertile plants from other recurrent populations and newly-added parents are all used as paternal line in plot I. The rest follows Program II.

The above programs have a common point in that selection and gene recombination are included in every generation at the same time, so that each generation can complete a cycle and obtain certain genetic gains.

Of these, Program I is the simplest one and is applicable for preflowering traits like cold drought and humid resistance, salt and alkali tolerance and vegetative growth habit. Besides making full use of environmental selection pressure, one should catch the keystage of the objective characters fully expressing and eliminate strictly the undesirable plants in the field without considering the fertility, thus the genetic base of both male and female gametes can be improved simultaneously.

Program II is applicable not only to preflowering characteristics but to the post-flowering ones, such as disease resistance at later growing stages, quality and many traits related to yields. Nevertheless, its selection accuracy is still not very high because the selected fertile plants giving the male gametes belong to F_1 generation.

Program III can offer more accurate selection of male gametes. Because F_1 selected fertile plants have had another generation of self pollination, some desirable recessive characteristics may be expressed and stabilized through selection.

Very good selection effect can be achieved with pollen from the best plants of the best rows. This program can select for almost every characteristic.

Program IV is merely adopted in a few cases.

These four progams should be applied actively and flexibly according to the objective characteristics and the degree of genetic variation in the recurrent population.

In applying program II and III, seeds of the selected sterile plants from plot I (except for a small amount kept for maternal lines in the next cycle) should be planted in plot II, so as to enlarge the selection population. Plot II and plot III can supply both pollinator and breeding materials. Their selection criteria should be some what different; the former demands predominantly good characters, while the latter requires excellent integrated characters. In certain cases, female gametes can be selected for high yield traits and male gametes for objective ones. If the objective trait is disease resistance, artificial infection should be induced. If the objective trait is quality, a primary selection should be conducted in the fields for the integrated characters or relevant ones, then each individual plant should be threshed and numbered. A certain amount of seeds are tested for quality. Relying on the results of evaluation, the remaining seeds having better quality will be blended and planted as the pollinator in the next cycle.

These programs are also suitable for other self-pollinated crops in applying dominant male sterile gene to recurrent selection.

DEVELOPMENT OF GENE POOL WITH IMPROVED RESISTANCE TO SCAB

Wu Zhaosu
(Nanjing Agricultural University)
Shen Qiuquan
(Hangzhou Agricultural Science Institute)
Lu Weizhong
(Jiangsu Academy of Agricultural Sciences)
Yang Zanlin
(Anhui Academy of Agricultural Sciences)

Introduction

Wheat scab is widespread on wheat in temperate humid and sub-humid regions of the world, and ranks second in importance of the wheat diseases in China [16]. It is the most destructive disease in the Central and Lower Yangtze Valley and the South China winter wheat regions, and in the east of the Northeast spring wheat region [1,16,21]. It is estimated that in China the disease has appeared in over 7 million hectars, about one-fourth of the total wheat acreage.

It was reported that, throughout the Central and Lower Yangtze Valley region during the 23 years between 1957-1979, severe epidemics occurred in 3 years and moderately severe ones in 12 years; in the severe epidemic years, the number of infected heads ranged from 50 to 100 percent and the yields were reduced by 10-40 percent; and in the moderately severe epidemic years, the number of infected heads ranged from 30 to 50 percent and yields were reduced by 5-15 percent[2]. This disease not only causes severe loss in yield and quality of grain, but also spoils the seed.

Progress has been made in breeding for scab resistance in China during the last 25 years, and yet the resistance has not been further improved and well incorporated into high yielding genotypes. Thus breeding for scab resistance has increased in emphasis in the overall breeding objectives in the above-mentioned regions, and has been included in objectives of the national wheat breeding project. Only limited progress has been made in other countries. Since 1982 CIMMYT has initiated new activities, including the development of source population (gene pool) for scab resistance[17].

This paper reviews the advances in the studies of the nature and inheritance of wheat scab resistance, and the progress in breeding for resistance in China; and with this data, it also discusses the breeding strategies, and presents a feasible scheme for using the single dominant male-sterile gene *Ta*1 in the development of a gene pool with improved resistance to scab.

Scab Resistance in Wheat

Wheat scab is caused by a number of *Fusarium* species, *F. graminearum* Schw. (*Gibberella zeae* Petch.) being the most prevalent in China. The causal fungi are facultative nonspecific parasites. Immunity and high degrees of resistance to scab have not been found in wheat and its relatives, but wheat varieties differ in their reactions significantly. Shroeder and Christensen (1963)[25] suggested two types of scab resistance in wheat: resistance to infection and resistance to fungal spread within the tissue after infection. Most Chinese workers have agreed with this viewpoint, and believe that the latter type should be the main component of resistance[3,4]. As for the mechanism of infection, no satisfactory conclusion has been drawn, though many observations have shown that infection mainly starts at the anthers. Recently Strange et al. (1978)[28] have confirmed that choline and bataine contained in the anther act as stimulants of *F. graminearum*, nevertheless the anther is not the only path of infection[28].

Only limited reports on inheritance of wheat scab resistance are available. A few reports indicate that the resistance is controlled by a few genes[14], while others suggest that it is controlled by multiple genes[25, 29]. In China, various workers have concluded that the scab resistance in wheat is quantitatively inherited, with low heritability and with some evidence of the additive effect of genes[5, 6, 9, 10].

Dai et al. (1941) studied extensively, the correlation between morphological characteristics and scab resistance[11]. Since then various conclusions have been drawn. It seems that some morphological characteristics may affect the expression of resistance under certain conditions, and there is no necessarily inherent relation between them. In general, high yielding, semidwarf wheat genotypes tend to be rather susceptible in China, and the same situation has been noticed abroad[29]. Recently, Gocho et al. (1983) reported that unfavorable genetic correlation, when attempting to breed a scab resistant variety having many desirable traits, was not found to occur between the scab resistance and the agronomic characteristics studied[19]. Thus it can be seen that a high yielding variety with improved resistance to scab could be developed by an appropriate method.

Progress in Breeding for Scab Resistance

During the last 25 years, considerable progress in breeding for scab resistance has been made in China. In the early stages, moderately resistant varieties have been developed from natural variants of the widely adapted, susceptible varieties by single plant selection, such as Wannian 2 and Wangmai 15 from Nanda 2419, and Wumai 1 from Funo; a moderately resistant variety has also been developed from the induced mutation of a susceptible variety, such as Emai 6 from Nanda 2419. The first scab resistant variety produced by hybridization was Sumai 3, which has been an excellent source of scab resistance through out the world. It is the result of a cross between two susceptible varieties, Funo and Taiwan-xiomai. Another two moderately resistant varieties, Huazhong 2133 and Zhengmai 7495, are also the results of crosses between two susceptible varieties. In addition, some moderately resistant varieties have been developed from hybridization involving rye as a parent, such as Jinzhou 1, 47 and 66.

On the basis of improved resistant varieties, further improvements have been achieved. Emai 9, originated as a plant selection from Emai 6, has yielded more than Emai 6. By using developed resistant varieties as parents in hybridization, more improved varieties have been developed. For example, Xiangmai 1 and Ganmai 359 are from different crosses involving Wannian 2 as a resistant parent and they yield more than Wannian 2. The Jiangsu Academy of Agricultural Science used Sumai 3 as a resistant parent in multiple crosses and developed the new lines Ning 7840, 8026 and 8017 with resistance to rusts and powdery mildew in addition to moderate resistance to scab and better agronomic traits than Sumai 3. Thus far, scab resistance has not been further improved and well incorporated into high yielding genotypes. The degree of resistance of almost all improved varieties does not exceed some indigenous varieties, such as Wangshuibai, Wenzhou-Hongheshang, etc.

For further improvement of breeding for scab resistance, supporting researches have been undertaken. Appropriate techniques for artificial inoculation and accurate indices of scab resistance have been developed[4,26]. Up to 1979, more than 10,000 wheat entries collected from various regions in China and from abroad have been screened for resistance at various institutions, over 400 moderately resistant and 10 higher resistant sources have been identified[12]. During the last few years, more entries and resistant sources have been screened and identified. Much information about the various resistant sources has been obtained.

Strategy in Scab Resistance Breeding

Progress in breeding for scab resistance in China has been greater than in other countries, yet this appears rather slow in contrast to the breeding for stripe rust resistance, early maturity and high yield. To solve this complicated problem, it is neccessary to use an appropriate breeding strategy.

In China the method used in the early stages actually appears to be no strategy at all. Whether the resistant varieties developed from natural variation, such as Wumai 1 and Wannian 2, or from induced mutation, such as Emai 6, these varieties did not result from conscious selection for the scab resistance. For the resistant varieties derived from hybridization between two susceptible parents, such as Sumai 3 and others, resistance had not been expected either. The latest approach to resistance breeding is by using a developed resistant variety as a parent once in hybridization, such as for the development of Xianamai 1, Ning 7840, etc. Scab resistance needs to be increased and incorporated into high yielding genotypes. Past strategie are not appropriate.

In order to improve scab resistance, yield capacity and other desired traits simultaneously, and to reduce the time required to complete breeding cycles to meet both the short- and long-term goals of the breeding program, the development of a gene pool with improved scab resistance and other desired traits should be the most appropriate strategy.

With quantitative inheritance, low heritability and multiple characteristics involved, the problem can only be effectively approached through parent building and recurrent selection[18]. Ramage (1977, 1980) emphasized that recurrent selection is concerned primarily with the generation of variation and is used to develop populations and information for exploitation by other breeding schemes[23, 24]; It is a breeding technique designed to accumulate or concentrate genes for a particular quantitative character without a significant loss of genetic variability, and is the most feasible means of simultaneous selection for a character and its most favorable background genotype.

Recurrent selection methods have been applied successfully in both cross- and self-pollinated crops for improving yield, quality, resistance, etc., and various schemes and procedures have been developed[20]. The applications of genetic male sterility to recurrent selection schemes in self-pollinated crops are attracting great interest. Athwal and Borlaug have outlined some techniques using recessive male sterile genes for the improvement of wheat populations[15]. Ramage has presented the technique using genetic recessive male sterility to facilitate recurrent selection in the breeding of barley and wheat[23]. In using a dominant male sterile allele in recurrent selection, Sorrells and Fritz have presented several schemes[27]. Deng and Ji have discussed the utilization of the single dominant male sterile gene $Ta1$ in recurrent selection in wheat breeding[13].

With increased facility in using the single dominant male sterile gene $Ta1$ and a number of valuable sources of various types of scab resistance, and with the increasing amount of genetic information presently available, a feasible scheme for the development of a gene pool with improved resistance to scab in wheat has been formulated, and parent-building has been initiated.

Establishment of Source Population

In order to create a useful germplasm population more efficiently, choosing parent materials becomes an important issue. Qualset and Vogt (1980)[22] suggested that several broadly adapted varieties may be chosen as the basic parents in crosses with a large sample of new source materials[22]. If the new source materials are known to have valuable new genes, but are grossly unadapted to regional conditions, they should first be used in a cycle of parent-building; the resulting materials would have better general adaptation and the loss of desirable genes would be less likely due to adverse combinations with undesirable genes. Previously, Athwal and Borlaug (1967)[15] noted that a broad based gene pool can be developed by making a collection of varieties and lines with a wide range of genetic diversity as the pollen parents.

The primary objective of this program is to develop a high-yielding gene pool well-adapted to the Central and Lower Yangtze Valley wheat region with improved resistnce to scab and other desired traits. Thus, the basic parents should be the various high-yielding cultivars currently grown in this region, and the majority of the crossing partners should be the stocks of various origins with a consistently higher degree of scab resistance, and the others should be the sources of resistance to other diseases and of other desired traits. Accordingly, 5 basic parents and 20 crossing partners have been chosen and used in parent-building.

5 basic parents with the single dominant male sterile gene $Ta1$ were each crossed with 20 partners, and considerable number of the crossed seeds of 100 combinations were obtained. In the next season, the F_1's of the 100 combinations will be grown in separate rows in an isolated block, in which each of the 20 combinations of the same basic parent will be planted in each of 5 sub- blocks, at each of 4 stations. Since progeny of every cross always segregate 1:1 sterile to fertile, natural crossing must be expected to take place mainly within individual hybrid populations in this arrangement. Mass selections of male sterile plants and of male fertile plants in each hybrid population will be made respectively, and then the seeds from selected male sterile plants and from selected male fertile plants within the same group (sub-block) will be composited separately. Lastly, each of 5 composite seeds from selected plants and 5 from selected male fertile plants of 5 groups of combinations from each station will be bulked respectively for the next trial.

In the next season, an isolated block with 5 sub-blocks will be set up at each station as before. In each sub-block, the bulked seed from selected male sterile plants of one group will be sown in alternate rows adjacent to rows of the bulked seeds from selected male fertile plants of the other four groups with the same proportion. All of the male fertile plants within the progeny of selected male sterile plants of each group will be rogued before anthesis, when they are easily identified, so as to assure random mating anong groups. Selection of male sterile plants throughout the isolated block, followed by compositing the seed from the selected plants, will be practiced.

The increase of the bulked intercrosses, which will contain genetic variability for the desired traits, will serve as the source population for the first cycle of recurrent selection.

Recurrent Selection Procedure

Population improvement methods fall naturally into two groups, those based on purely phenotypic mass selection and those based on selection with progeny testing[26]. Phenotypic mass selection seems to be effective for most traits in most crops, and can be especially effective for improvement in populations into which exotic or unadapted germplasm has been introgressed[20]. This method is advantageous in shortening the time between cycles of recurrent selection, and is more likely to maintain genetic variability to permit continued improvement. But for less heritable traits, recessive traits, and those complex traits whose expression cannot be easily identified among individual plants, selection with progeny testing should be more effective than phenotypic mass selection. In this program, some of the donor parents used are grossly unadapted to the agroecological condition of this region, and the others, such as indigenous varieties, are not adapted to higher levels of production, phenotypic mass selection should be practiced primarily for improved adaptiveness and the maintenance of genetic variability in the improved population. However, in considering the critical trait, such as scab resistance, phenotypic mass selection seems to be less effective. Scab resistance is of low heritability, and narrow range of variability, and is biased by male sterility. The male sterile plants usually appear to be less susceptible because of their contracted anthers, which tend to obscure the genetic effect. The current expression of individual male fertile plants does not give reliable evidence about the genetic effect of resistance, which would be better identified through progeny testing. Based on this test, selections of desirable plants from resistant families could be used as pollinators in the intercrossing population to promote further improvement, and the superior genotypes could be extracted to develop new cultivars. In short, this gene pool is to be developed on the basis of multiple-parent crossing and through recurrent selection by integration of phenotypic mass selection and progenytesting selection.

The procedures are outlined.

The recurrent selection program begins with the establishment of the source population. From this population, the male serile plants and the male fertile plants are selected respectively. With the male sterile plants, phenotypic mass selection with single spike descent method instead of single-seed descent method[22] is exercised for agronomic characters and general resistance to rusts and powdery mildew. With the male fertile plants individual plant selection is conducted for additional traits, such as yield components and special resistance to scab. The bulked seed of selected male sterile spikes serves as the base population for the next cycle of recurrent selection. The seeds of selected male fertile individual plants serve for progeny testing.

In the second cycle of recurrent selection, the above procedures are repeated in the base population. The bulked seed of selected male-sterile spikes serves as the base population for the third cycle of recurrent selection, and the seeds of selected male fertile individual plants serve for progeny testing.

Besides, the progenies of the selected male fertile plants are grown in a plant-to-row designed nursery, in which evaluation and selection for desirable characters involving scab resistance are imposed both among and within plant rows. Individual superior plants within selected rows are selected, and the seeds of selected plants are partially composited to serve as pollinators in the next intercrossing block, and the rest seeds carry on in the conventional breeding program.

In the third cycle of recurrent selection, the base population is planted in alternating rows with the pollinator selected from progeny testing nursery in adjacent rows, so as to increase the frequencies of desired genes in the next hybrid population derived from the male sterile plants in this bass population. Phenotypic mass selection of male sterile plants and individual plant selection of male fertile plants is practiced from this base population, and other procedures including progeny testing are repeated as mentioned above.

Following cycles of recurrent selection are repeated for further improvement of population.

For continuing development of the gene pool, it is necessary to introduce additional sources of germplasm in any cycle. To avoid bringing in undesirable genes, these new germplasm sources must be carefully chosen and crossed or backcrossed onto male sterile plants selected from the improved population, and the male fertile plants with the desired genes selected from progenies of the crosses could be used as pollinators in the intercrossing populations.

In this program, a cycle of recurrent selection is mainly completed in one year, although the complementary selection with progeny testing requires two years. After the second cycle of phenotypic mass selection, the pollinator for the intercrossing population is provided by selection with progeny testing every year. In this way, the hybrid population could be improved every year, and the opportunities to extract superior genotypes are enhanced. Both isolated intercrossing block and progeny testing nursery are inoculated with scab fungi to facilitate selection for resistance. From the very beginning of this program, large scale experiments and multi-location tests are kept to assure the success of this program.

Organization and Coordination

This research is complex, and cannot be fulfilled by individual institutions working in isolation. The success of this task lies entirely in close cooperation among participating institutions and scientists from related disciplines under a unified program. The features of this program include the use of the single dominant male sterile gene $Ta1$ as genetic tool, multiple-parent crossing to establish broad based source population, recurrent selection by integration of phenotypic mass selection and progeny testing selection to accumlate various desirable genes, especially of resistance to scab, in the improved population; and in any cycle, new sources to be incorporated into the population to permit continued improvement, and superior genotypes to be extracted from the population to develop new cultivars. This gene pool is to be developed in a continuous process to meet both the short- and long-term goals of the breeding programs in this wide region.

Much work needs to be done by large-scale cooperation on the solution of the problems mentioned through the program, such as systematic evaluation and effective utilization of various sources of germplasm, establishment of source population with broad genetic base and available genetic variability, efficient techniques of development and exploitation of the gene pool, and assessment of yield potenţial, yield stability and adaptability of promising materials, and so forth.

The establishment of inter-institutional research team in the Central and Lower Yangtze Valley region has been initiated. Source population establishment activities are under way. The participating institutions have worked under the unified program, and the germplasm and materials generated at each station have screened through multi-location tests. Through mutual visits to different institutions and discussions, analysis of results on the regional level, gathering the desirable materials generated at each station and redistribution to the participating institutions, through reviewing the previous work and planning the program for the following year and other activities, this program will be carried out effectively and the progress of breeding for scab resistance will be accelerated.

It is certain that if an effective interdisciplinary research task force is established to continue and strengthen close cooperation among participating institutions and with improved programs, a breakthrough in wheat improvement in this region and the related regions will be achieved.

References

[1] 金善宝,1983,中国小麦品种及其系谱,农业出版社。
[2] 李克昌,1982,小麦赤霉病及其防治,上海科技出版社。
[3] 王裕中、杨新宁、肖庆璞,1982,小麦赤霉病抗性鉴定技术的改进及其抗源的开拓,中国农业科学,(5):67-77。
[4] 徐雍皋、方达中,1982,玉蜀 赤霉对小麦品种致病力的测定方法和致病力分析,植物病理学报,12(4):53-57。
[5] 赵忠贞、闵捷,1980,小麦品种抗赤霉病遗传性的初步观察,浙江农业科学,(6):286-287。
[6] 尹腾蛟、尹世强,1981,小麦籽粒对赤霉病抗性的遗传力的初步研究,湖南农业科学,(5):31-35。
[7] 张乐庆、潘雪萍,1982,小麦品种对赤霉病抗扩展性的遗传研究,华南农学院学报,3(4):21-29。
[8] 余毓君,1982,小麦品种苏麦3号抗赤霉病性及产量因素的单位分析,华中农学院学报,(2):70-77。
[9] 陈楚和,1983,小麦抗赤霉病遗传的研究,浙江农业大学学报,9(2):115-126。
[10] 顾佳清,1983,小麦赤霉病抗性遗传的研究,中国农业科学,(6):61-64。
[11] 戴松恩、姜权、王焕如,1941,小麦 种抗赤霉病之育种问题,农报,(6):616-625。
[12] 全国稻、麦、玉米、棉花抗病育种学术讨论会纪要,1979。
[13] 邓景扬、纪凤高,1983,显性雄性不育核基因Tal在小麦育种上的价值与主要利用途径,中国农业科学,(4):6-11。
[14] 中川元兴,1955,小麦品种赤"病抵抗性研究,第二报,赤力"病耐性遗传因子,15-22。
[15] Athwal, D.S. and N.E. Borlaug. 1967. Genetic male sterility in wheat breeding. *Ind. Jour. Genetics and Plant Breeding,* **27**(1): 136-142.
[16] Chiu, W.F. and Y.H. Chang. 1982. Advances of science of plant protection in the People's Republic of China. *Ann. Rev. Phytopathology,* **20**: 71-92.
[17] CIMMYT Review, 1982, El Batan, Mexico.
[18] Everson,, E.H. and C.R. Olien., 1975, Breeding wheat for winter hardiness. Proc. 2nd Inter. Winter Wheat Conf., 74-83.
[19] Gocho, H. et al. 1983. Induced mutations and genetic correlations between scab resistance and some agronomic characters in wheat. Abs., Proc. 6th Int. Wheat Genet. Symp., Kyoto, Japan.
[20] Hallauer, A.R. 1981, Selection and breeding methods. In Frey, K.J. (ed.), Plant Breeding II. Iowa State Univ. Press. 3-56.
[21] Hanson, H., N.E. Borlaug, and R.G. Anderson, 1982. Wheat in the Third World. Westview Press, Inc.
[22] Qualset, C.O. and H.E. Vogt, 1980, Efficient methods of population mangement and utilization in breeding wheat for Mediterraneantype climates. Proc. 3rd Int. Wheat Conf. 166-188.
[23] Ramage, R.T., 1977, Varietal improvement of wheat through male sterile facilitated recurrent selection. ASPAC Tech. Bull. No. 37.
[24] Ramage, R.T., 1980, Genetic methods to breed salt tolerance in plants. In Rains, D.W. et al. (ed.) Genetic Engineering of Osmoregulation. Plenum Publishing Corp. 311-318.
[25] Shroeder, H.W. and J. J. Christensen, 1963, Factors affecting resistance of Wheat to

scab caused by *Gibberella zeae*. *Phytopathology,* **53**: 831–838.
[26] Simmonds, N. W., 1979, Principles of Crop Improvement,, Longman.
[27] Sorrells, M. E. and S. E. Fritz, 1982, Application of a dominant male-sterile allele to improvement of self-pollinated crops. *Crop Sci.,* **5**: 1033–1035.
[28] Strange, R. N. and H. Smith. 1978. Specificity of choline and betaine as stimulants of Fusarium graminearum. *Trans. Brit. Mycol. Soc.,* **70**(2): 187–192.
[29] Tomasovic, S. et al., 1983, Breeding and studying wheat for resistance to *Fusarium* sp. especially to *Fusarium graminearum* Schw. (Petch.). *Ann. Wheat Newsletter,* **39**: 162–164.

A RECURRENT SELECTION SCHEME FOR BREEDING WHEAT VARIETY WITH RESISTANCE TO SCAB (*Gibberella zeae* (Schw.) Petch) COLONIZATION

Zhang Leqing

(*South China Agricultural University, Guangzhou, China*)

The objective is combine many resistant genes into a mixed community by using agronomic parents with the best genetic background and dominant genic sterile genes as basic materials and by recurrent selection[3, 4]. After inoculation and selection, recurrent intercrossing and recombination, an improved community may be formed, from which a mixed line or pure line with high resistance may be obtained. The steps of the procedure involve polymeric hybridization (the formation of basic community), recurrent selection and utilization of the resistant products (Fig. 1).

Outline of Methods Involved

Polymeric Hybridization (Formation of the Basic Community)

1. The genic sterile materials from Fu-Fan No.17, a popular variety, are used as maternal parents for crossing with eight lines with resistance to wheat scab, such as 12 G-8-5, Fan No. 60085, Fan 60096, Fan No. 538, Fan 635, Fan No.3, Fan 50002 and Fan 50006 etc.which possess high resistant characteristics and which come from different resistant resources.

2. The seeds from eight crossing groups are mixed and planted under separate conditions. They are allowed to cross with each other adequately. This procedure is carried out over three successive generations. The community of the third intercrossing generation finally consisting of about 30,000 individual plants. The wheat grain cultures are put on the soil surface to induce the outbreack of wheat scab in the third intercrossing generation.

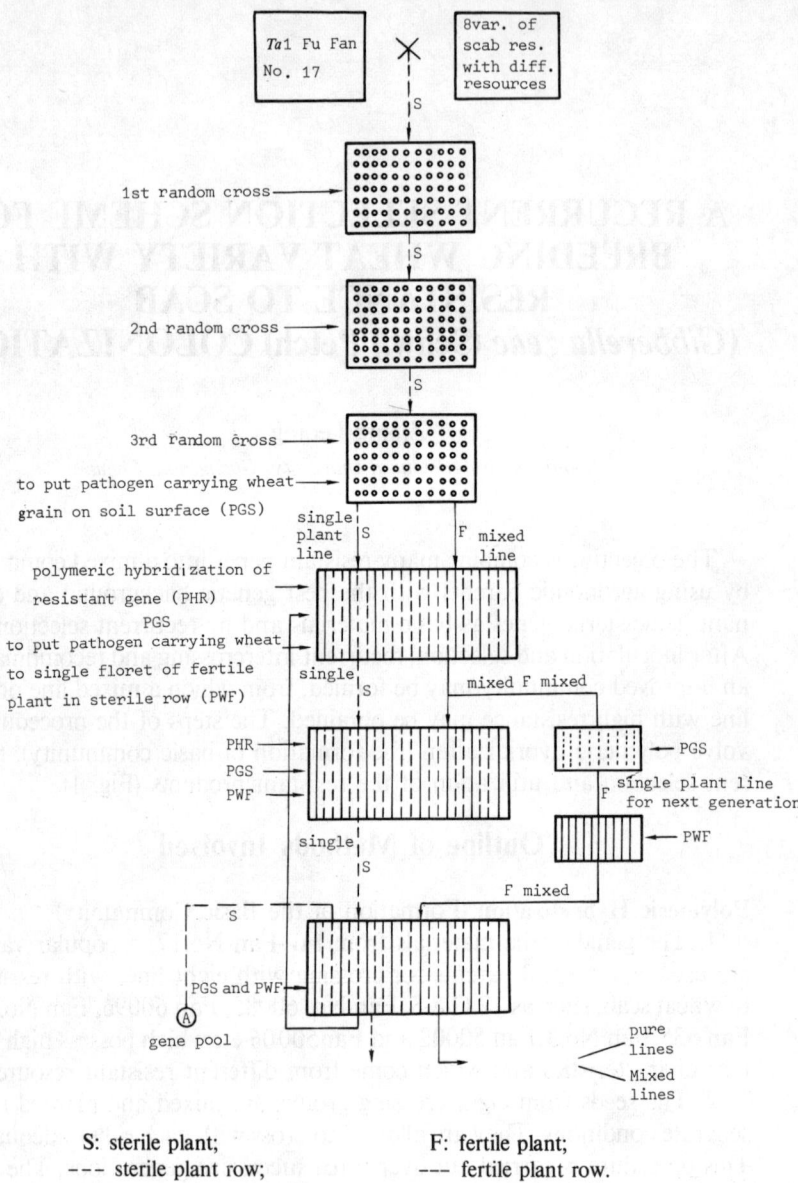

Fig. 1 The working plan for recurrent selection of wheat varieties with resistance to scab colonization.

Then, the fertile plants with higher resistance are chosen and the seeds are mixed as male parents in F_1. The sterile plants which are susceptible to the disease are eliminated. About 20 per cent of sterile plants with less infection are planted in one row between rows of mixed fertile parents in the following year to create a basic community.

Recurrent Selection

From the maternal parent rows, twenty per cent of plant rows are chosen. In each row, three to five sterile plants are selected and planted still in row as maternal parents following year. Another ten fertile plants are selected. Half of them are used as male parents and planted in row between maternal parent rows in following year to form a first recurrent intercrossing community. The other half of them are taken to separate plots for the test of soil surface inoculation. The plants with resistance are selected and planted for one more generation to perform single flower injecting infection. The plants with resistance are chosen from the resistance lines and the seeds are mixed to serve as male parents in the third recurrent.

The second and third recurrent: The treatment is the same as the first recurrent. About twenty per cent of good lines are chosen from maternal parent lines each recurrent and the fertile plant are gradually introduced to serve as male parents from the good fertile lines which have been identified in past generation.

The main link of inoculation and identification is that wheat grain cultures are put on soil surface as inoculum of scab in recurrent community. Sterile row with serious disease are eliminated. At the same time, good fertile plants in sterile rows are inoculated by means of injecting one drop of spore suspension with a small injector to one floret each ear. Bags are used to separate the plants from each another, the fertile plants with high resistance are thus selected. The same amount of seeds from each plant is mixed. Half of them are used as male parents next year. The other half of them are multiplied to produce F_2 in north China in the same year. After an artificial inoculation on soil surface, individual plants with little infection are selected. Seeds from one plant are sown in one row at the original breeding place for F_3. After an artificial inoculation by means of injecting the spore suspension to the floret of the vigorous single plant rows in the community, look for rows showing less infection in spikelets, no infection in ear axis and high resistance to colonization. The seeds from individual plants in such rows are mixed as male parents and planted among sterile plant rows in the following year. This method could raise the hereditary progress of resistance of each recurrent and form animproved community.

The Utilization of the Improved Community

About 1000–1500 fertile plants are chosen from the good sterile rows. They are planted in single plant rows and transfered to conventional breeding procedure. Thus pure line and mixed variety which have transgressive resistance are obtained. The seeds from sterile plants can be used as a gene pool of scab resistance.

Several Key Techniques in Recurrent Selection Method for Wheat Scab Colonization Resistance Varieties

The Selection of Parent Material

The resistance to scab colonization is controlled by multigene. The nature of the recurrent selection is to improve the resistance by exploiting the additive effect of gene[2]. Hence, it has been emphasized that the resistance parents employed in the working plan come from different resistant resources. For example, the resistance of Fan No. 538 comes from Fujiam and Shangmai which are local varieties of Fujien Province. 12 G-8-5 and Fan 635 comes from Fan-shan, Fan 60085 and Fan No. 60096 are the combination product of Sumai No. 2 and Zhenzhou No.1. The resistance of parent materials must be not only strong but also stable enough. The resistance of the parents, especially those of Fan No. 50002, Fan No. 60085 and Fan No. 60096 in the working plans are higher than those of Sumai No. 3 in the coordinate test in five provinces and cities of China where wheat scab often occures.

The Selection of Fertile Plants in Recurrent Selection for Wheat Scab Colonization Resistance Should Be Emphasised

Wheat scab occures during the flowering period. In general, the pathogen infects anther first, and then spreads. Sterile plants have small, thin and infertile anthers. They usually remain within the hull. Hence, the sterile plants are less infected than fertile plants within a community. It is not reliable to choose them according only to the appearance of sterile plants. Because the flowers of fertile plants may develop normally, the degree of infection here will more truly reflects the degree of resistance. Since the average genetic components in the sterile progeny derived from the resistant materials containing Ta1 sterile gene are the same as those in fertile progeny, only those infected in sterile plants or lines are eliminated under the pressure of artificial inoculation in the working plan. At the same plot,, inoculation method of injecting the spore suspension to florets is used to test and choose high resistance of individual in fertile plants. In this way, the improvement of resistance of fertile plants is expected. They are then used as pollen donor in next progeny.

The Inoculation and Identification Technique in Recurrent Selection

In the early 60's, Schroeder divided resistance into entrance resistance and colonization resistance[6]. Because all resistant materials belong to colonization resistance at present, the requirements for pollen donor are restricted to such a extent as small portion of infected spikelet, the slow colonization of the disease, and little loss of the yield. Special inoculation and identification techniques must be used to determine the colonization resistance. The soil surface infestation method and the floret injecting inoculation method[5] are the two methods usually applied.

The soil surface infestation method is suitable for the inoculation of a big community in order to discharge those infected sterile plants and fertile plants. Whether the materials passing through such inoculation test have real scab resistance is not sure. Even though they have real resistance, it is hard to measure the degree of colonization resistance.

The floret injecting inoculation method is suitable for plots with single plant row. Fertile plants with good agronomic characters in sterile row of the plot will be chosen. When the first flower of a floret in the middle of these ears is coming on, such inoculation procedure is carried on. It is possible to decide whether a sterile plant should be discharged or selected according to the degree of colonization and the infection of ear axis.

The Utilization of the Improved Community

As it is known that the popular varieties with a narrow basis of inheritance are vulnerable to disease, Marshall proposed a scheme to improve the disease resistance by increasing their genetic diversity[1]. Regardless of the result of the controversy about whether the pathogen of wheat scab can differentiate[1], the heterogeneity of a community can slow down the spread of disease and stabilize the disease resistance of variety. It also helps to make the yield stable and to raise the adaptability of crop community. Hence, the choice of highly resistant fertile plants provides not only a base for pure varieties, but also mixed varieties. the latter is even more important. Individual plant lines which have similar morphological and biological traits are mixed and can be used in field production.

References

[1] 李克昌，1982，小麦赤霉病及其防治，上海科学技术出版社。
[2] 张乐庆、潘雪萍，1982，小麦品种对赤霉病的扩展传的遗传研究，华南农学院学报，3（4）：21-29。
[3] 邓景扬、高忠丽，1980，小麦显性雄性不育基因的发现—太谷不育小麦鉴定总结，作物学报，6（2）：85-98。
[4] 邓景扬、纪凤高，1983，显性雄性不育基因Tal在小麦育种上的价值与主要利用途径，中国农业科学（4）：6-11。
[5] 福建农学院作物教研室，植物病理教研室，1978，小麦抗赤霉育种工作的初浅认识，福建农业科技（1）：2-4。
[6] Schroeder. H.W. amd J.J. Christensen, 1963,, *Phytopathology*, 53: 831-838.
[7] Marshall. D.R. 1977, The advantages and Hazards of Genetic homogeneity. In the Genetic Basis of Epidemis in Agriculture. pp. 1-20. (ed by P.R. Day.) Annals of the New York Academmy of Sciences, New York.

IN THE MULTIPLE RESISTANCE BREEDING

Lin Zuoji

(*The Wheat Institute, Henan Academy of Agriculture and Forestry Sciences*)

Wang Zhenfu and Shuang Zhifu

(*The Crop Genetic Institute Shanxi Academy of Agricultural Sciences*)

In recent years, more and more attention has been paid to the breeding of multiple resistant varieties (resistant to many pathogens and multiple races of one pathogen) with advances in plant breeding and the improvements in cultivation techniques[4, 6, 8, 10]. In many countries and regions, quite a few varieties resistant to rusts and powdery mildew have been bred, such as Zorba, Salmünder Bartweizen, Kavkas, Lovrin-10, Veery"s", T.J.B series and Fengkang series. etc. With the discovery of the dominant genic male sterile wheat in Taigu[1, 2], we assume that, given the characteristics of this material that by comprehensively utilizing the methods of backcrossing, convergent crossing and recurrent selecting, multiple resistant genes would be assembled. Then through thoroughly recombining populations of larger sizes, the chances of selecting both disease resistance and the best background genotypes would be increased to favour the developing of multiple resistant composite varieties and pure lines[3, 5, 7, 9, 11].

The Procedures to Improve Disease Resistance of Common Varieties (Program I)

Crossing and backcrossing

The recurrent parents must be chosen from those with the best background genotypes. At present, the stable male sterile material derived from some released varieties or new lines with high yield potential and higher general combining ability might be used. The donor parents should be selected on the disease resistance identification basis from: 1) material with strong and stable resistance to the main pathogens and their main physiological races. Also material with resistant genes from different resources and those donor parents resistant to multiple main pathogens would be all the better; 2) material with dominant resistant genes and strong hereditary capacity of resistance; 3) material without serious defects but with favourable characteristics which complemented with those of the recurrent parents. The number of the parents with the resistant genes depends on the species of the pathogens and the number of the races. For the main pathogen, one or two more may be used, but the total is not to go beyond 15. Backcrossing is to be done in different disease nurseries (such as yellow rust, leaf rust and powdery mildew nurseries in our province).

Convergent crossing

Some combinations should be eliminated according to the performance of the backcross progenies. Then, in accordance with the main diseases the parents are resistant to, the seeds of the backcross progenies from the sterile plants are to be mixed in groups and mated in diallel crossing. For example, sow the mixed seeds of AY (resistant to stripe rust) as female rows and those of AL (resistant to leaf rust) in adjacent rows as male. And, similarly, plant the seeds of AL/AP (resistant to powdery mildew) and those of AP/AY in isolated plots respectively. The male parents had better be sown at different times. At heading, all the fertile plants in the female rows must be removed to favour the recombinations of the genes, (partial reciprocal cross by hand pollinating might be carried out if necessary). Next year, random mating is to be done within the population composed of the mixed seeds from the female plants of all the three groups, and only those plants with the most unfavourable agronomic characters should be eliminated before flowering to prevent pollution of the population (Fig. 1).

Identifying in different disease nurseries

The seeds of the last generation,, having been mixed, should be divided into two parts and sown in isolated disease nurseries. The seeds from both the fertile and sterile plants of the last generation may be sown either mixed or in adjacent rows. Inoculate the main races of the locality in the fields, stripe rust and leaf rust in one nursery at different times and powdery mildew in the other, and harvest the seeds of the resistant fertile and sterile plants seperately.

Reciprocal mating among populations

The mixed seeds of the sterile plants from the rust nursery of last year should be sown as female rows in the same nursery and those of the fertile plants from the powdery mildew nursery of last year in adjacent rows as male. The seeds of the sterile plants from the powdery mildew nursery of last year should be sown in the same nursery as female and those of the

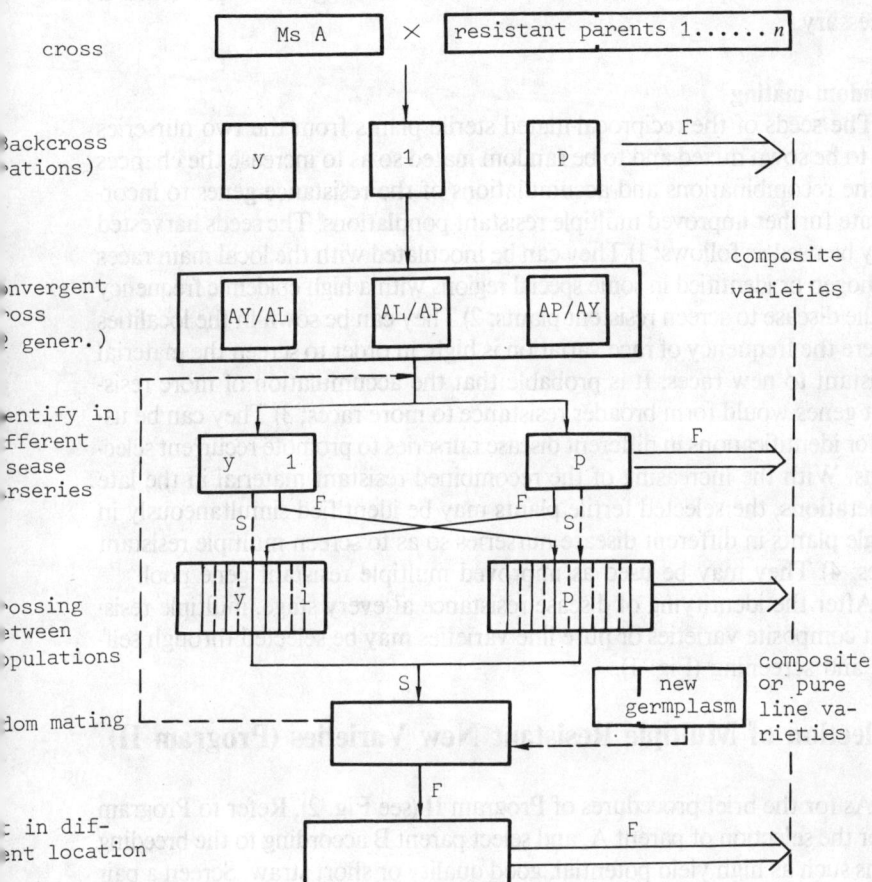

Fig. 1 The procedure of multiple resistant breeding emphasizing the improvement of resistance

S sterile plants;   y, 1, p inoculated mixed races of different pathogen;
F fertile plants;   ⇒ selfing and screening.

fertile plants from the rust nursery as male. At heading pull out the fertile plants of the female rows in both the nurseries to promote reciprocal mating among the populations, and at maturity harvest the resistant sterile plants in the female rows and the fertile plants in the male rows separately. Partial resistant fertile plants in female rows could be bagged and preserved if necessary.

Random mating

The seeds of the reciprocal-mated sterile plants from the two nurseries are to be sown mixed and to be random mated so as to increase the chances of the recombinations and accumulations of the resistance genes to incorporate further improved multiple resistant populations. The seeds harvested may be used as follows: 1) They can be inoculated with the local main races pathogen or identified in some special regions with a high epidemic frequency of the disease to screen resistent plants; 2) They can be sown in the localities where the frequency of race variation is high, in order to screen the material resistant to new races. It is probable that the accumulation of more resistant genes would form broader resistance to more races; 3) They can be used for identifications in different disease nurseries to promote recurrent selections. With the increasing of the recombined resistant material in the late generations, the selected fertile plants may be identified simultaneously in single plants in different disease nurseries so as to screen multiple resistant lines; 4) They may be used as improved multiple resistant gene pool.

After the identifying of disease resistance at every stage, multiple resistant composite varieties or pure line varieties may be selected through selfing and screening (Fig. 1).

Selection of Multiple Resistant New Varieties (Program II)

As for the brief procedures of Program II (see Fig. 2). Refer to Program I for the selection of parent A, and select parent B according to the breeding aims such as high yield potential, good quality or short straw. Screen a pair of parents from quite a few parents available through combining ability tests. The parents from 1 to i and those from 1 to j are all disease resistant and the two sets of resistant parents may be different in part according to the difference between A and B. The backcrossing procedure might be omitted (The backcrossed material of Program I may be utilized if the two programs are to be carried out simultaneously). The convergent crossing and the identifying in the different disease nurseries are the same as those in Program I. Then carry out reciprocal matings among different populations in different ways. For example, population AR derived from parent A and rust resis-

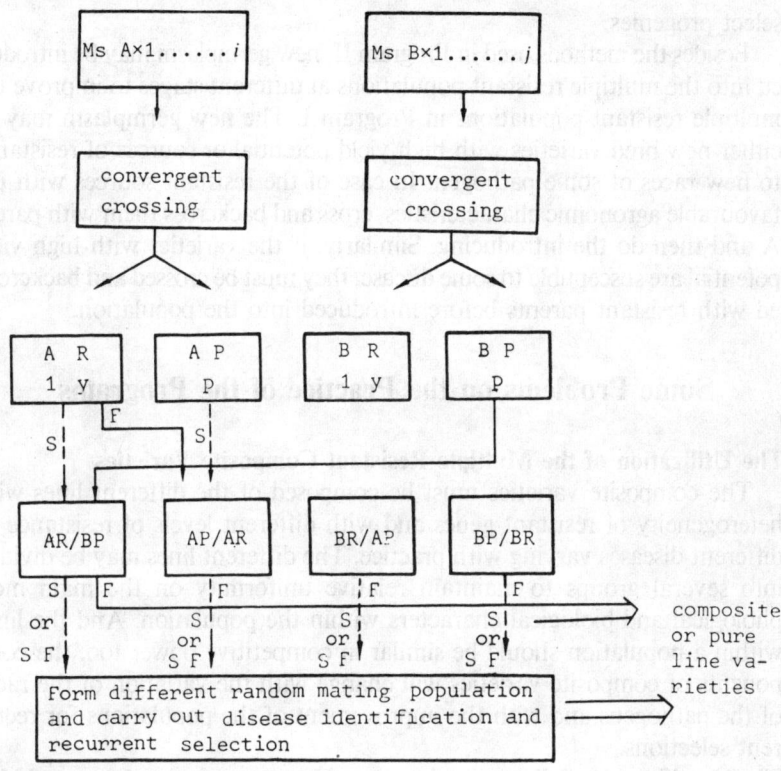

Fig. 2 The breeding procedure for the improvement of disease resistance, yield potential and other agronomic characteristics.

tant may be mated with population AP or population BP, derived from parent A or parent B and powdery mildew resistant. The random mating population recombinations of the next generation also depend on the breeding aims. For example, when parent A is as important as parent B, AR/BP//BR/AP may be used; while parent A is more important than parent B, recombine the populations as AR/BP//AP/AR. Population A and population B can be also improved in the direction A and direction B respectively. Refer to Program I for other identifying and recurrent selecting procedures only, aside from disease resistance, pay attention to the identifying and selecting of agronomic characteristics and use proper methods to identify and select progenies.

Besides the methods used in Program II, new germplasm may be introduced into the multiple resistant populations at different stages to improve the multiple resistant populations in Program I. The new germplasm may be either new bred varieties with high yield potential or sources of resistance to new races of some pathogen. In case of the resistant sources with unfavourable agronomic characteristics, cross and backcross them with parent A and then do the introducing. Similarly, if the varieties with high yield potential are susceptible to some disease, they must be crossed and backcrossed with resistant parents before introduced into the population.

Some Problems on the Practice of the Programs

The Utilization of the Multiple Resistant Composite Varieties

The composite varieties must be composed of the different lines with heterogeneity of resistant genes and with different levels of resistance to different diseases varying with practice. The different lines may be divided into several groups to maintain relative uniformity on the main morphological and biological characters within the population. And the lines within a population should be similar in competitive power too. the components of composite varieties will change with the variation of the races of the pathogens and with the improvement of the populations for recurrent selections.

The identification of disease resistance and preservation and accumulation of resistant genes

Identify disease resistance by inoculating the local epidemic races in the fields. Stripe rust and leaf rust may be inoculated in the same nursery at different times, with the population appropriately larger. Since the resistance to stripe rust is more important than that to leaf rust in our province. The plants resistant to leaf rust must be selected from those resistant to stripe rust. Select the resistant plants twice, first at heading stage and then at milky stage, those plants with leaf rust coming late and low percentage of severity also can be preserved. The identification may also be done in different natural infection sites and then the mating populations will be composed of the material (both the sterile and fertile plants) selected from all the sites. In general, resistance to all diseases is acceptable at lower level in the earlier generations for recurrent selections. Those plants with middle or slightly higher resistance and with wild genetic variation in agronomic characteristics may be selected, and the selection pressure will be increased gradually at later generations. The fertile plants with different types of resistant genes selected from those in different generations and sites may be preserved and, if necessary, returned to the multiple resistant populations so as to intensify the resistance to some disease or to prevent genetic shifts. In order to make full use of multiple resistant populations, material with minor genes resistant to some disease such as stripe rust may be selected by inoculating under different conditions such as different temperature and used for recurrent selection programs with lower selection pressure.

References

[1] 王琳清等，1980，小麦显性单基因控制的雄性不育材料"2-2-3"的研究及其利用，中国农业科学，(3)：1-7。

[2] 邓景扬，高忠丽，1980，小麦显性雄性不育基因的发现与利用—太谷不育小麦鉴定总结，作物学报，(2)：85-98。

[3] 邓景扬、纪凤高，1983，显性雄性不育核基因Tal在小麦育种上的利用价值与主要利用途径，中国农业科学，(4)：6-11。

[4] Borlaug, R.I. 1959. The use of multilineal or composite varieties to control airborn epidemic diseases of self-pollinated crop plants. Proc. 1st International Wheat Gentics Symposium, 12-27.

[5] Driscoll, C.J. 1983. The use of Cornerstone male-sterility in wheat breeding (Abst) 6th Intern. Wheat gentics symposium.

[6] Gautom, P.L. et al. 1979. synthesis of wheat multilines. Indian Journal of Genetics and Plant Breeding 39(1): 72-77.

[7] Jensen, N.F. 1970. A diallel selective mating system for cereal breeding. Crop Sci. 10(6): 629-635.

[8] Khush, G.S. 1978. Breeding methods and procedures employed at IRRI for developing rice germplasm with multiple resistance to diseases and insects. Tropical Agri Res. Series No. 11. Symposium on Methods of Crop Breeding.

[9] Krupinsky, J.M. and E.L.Sharp. 1979. Reselection for improved resistance of wheat to stripe rust. Phytopathology 69:400-404.

[10] Rajaram, S. et al. 1979. Diversity of rust resistance of the CIMMYT multiline composite, its yield potential and utilization. Indian Journal of Genetics and Plant Breeding 39(1)60-71.

[11] Sharp, E.L. 1979. Male sterile facilitated reccurent selection populations for developing broad-based resistance by major and minor effect genes. 9th Ihternational Congress of plan t protection Washingtop D.C. Abstract No 714.

THE RECURRENT SELECTION SCHEME FOR IMPROVEMENT OF MULTIPLE CHARACTERS

Shen Qiuquan

(*Hangzhou Agricultural Science Institute*)

Recurrent selection involves repeated intercrossing among many parents, thus it can maintain wide genetic variability in the breeding population, break down linkage and facilitate the accumulation of favorable alleles. It is also suitable for improving quantitative characters and setting up germplasm pools with wide genetic basis. Therefore, many breeders are interested in it and disirable results have been obtained in breeding programs of field crops[2, 5-9]. With the discovery of male sterile plants in wheat, a variety of schemes facilitating recurrent selection with male sterility have been developed[3, 4]. Most of them were aimed at single character improvement[10-13], but it is possible in many breeding programs to realize a comprehensive improvement of several target characters simultaneously. In this paper, a scheme for varietal improvement of multiple characters in wheat is presented. The breeding scheme for recurrent selection of polypopulation combination uses $Ta1$ male sterile dominant gene[1].

This scheme requires forming a synthetic population with multiple traits and several sub-populations with a particular single trait. Since the synthetic population contains parents with different trait types, it is difficult to achieve a improvement for a particular trait quickly in the process of recurrent selection for multiple traits. Meanwhile, a sub-population composed of a small quantity of germplasm of a single particular trait so it has more opportunities to accumulate the favorable alleles. Therefore, individuals showing transgressive inheritance are more likely to appear. By doing recurrent selection in a synthetic population and in sub-populations separately and the interchange of genes between the populations, the recombination of the favorable alleles controlling target traits in the synthetic population would take place. Genetic elements of key traits in the synthetic population are reinforced through a continuing input of the elite trait germplasm from branch populations. For improving multiple traits, the method of recurrent selection of the polypopulation combination would be more effecient than the recurrent selection carrying synthetic population alone. And this method requires fewer breeding years than the procedure which calls for the recurrent selection of individual population with different traits at first and then the synthetic recurrent selection. But it is not necessary to establish different sub-populations for each target traits, only for those important target traits having lower inheritability and being difficult to identify and with a poorer genetic base.

At first, the parents possessing the primary target traits and ecological adaptability should be in high proportion. All the parents should combine target traits and favorable ground genotypes with fairly large genetic differences. The parents should be classified into several groups according to their trait types. In each group there should be several $Ta1$ gene-carrying parents used as female parents for separate crosses with the other parents. The sterile F_1 hybrid is crossed with the male parents once or twice, then the seeds obtained are used to establish various populations. Synthetic population is formed by mixing hybrid seeds from all combinations; while subpopulations are formed from hybrid seeds from combinations having a certain target trait. Before the recurrent selection, several cycles of convergent crossing should be done and undesirable individuals should be discarded to improve the base population. Then recurrent selection is carried out separately using the same method for synthetic population and sub-populations, while they are closely connected through germplasm exchange between populations.

The recurrent selection for improvement of multiple characters involves both the mass selection in the present generation and the evaluation selection of progenies. Mass selection can be done in successive years with a shorter cycle of recurrent selection. But this kind of selection is limited to the phenotype of the male-sterile plants, therefore a lower selection pressure is needed. The evaluation selection of progenies involves selection among lines and within lines. It should be carried out in specific conditions or in certain generations which allow for identifications of different traits with a higher selection accuracy. But the number of families identified is so small that this would lead to a low genetic variability in the population. Combining these two selection methods could maintain the genetic variability of the population, since a population could be reestablished by using large quantities of male-sterile plants derived from mass selection under a lower selection pressure. It will also rapidly increase the frequency of favorable alleles in the population through the yearly crosses with the elite plants obtained from strict progeny evaluation selection; thus average performance of the population can be improved.

The different populations obtained from recurrent selection are closely combined through germplasm exchange among populations in the scheme of poly-population combination recurrent selection. This exchange primarily involves continuously bringing elite plants from the sub-populations into the synthetic population to supplement and improve key traits. The germplasm output of sub-populations strengthened by elite germplasm would also help increasing varied types in the synthetic population. Germplasm exchanges among populations should be carried out continuously during the process of recurrent selection. Concretely, the necessity of germplasm exchanges and the quantity of the germplasm for the exchange are to be determined according to the extent of difference between populations for a particular trait average level. The male-fertile plants selected through the progeny evaluation of the sub-population can usually be utilized as the germplasm for exchange and the male-sterile plants gained from mass selection in the present generation can also be used.

At each phase of whole recurrent selection, breeding materials can be obtained either from the recurrent selection populations or from the trait-evaluation plots or yield-test plots of the progenies. Using these materials pure-line varieties or trait germplasm can be developed by conventional methods. Besides, when population are reformed, the surplus hybrid seeds can be sowed in different areas for cooperative selections so as to raise the utility of population. Finally, the synthetic population would become a composite gene pool through continuously input of new germplasm and recurrent selection for several years. As the pool converges a large number of favorable alleles controlling the target traits for the local breeding program, the trait sub-populations would become gene pools with specific traits. Both type of gene pools would make contributions to the long-term breeding program.

References

[1] 邓景扬等, 1980, 小麦显性雄性不育基因的发现和利用 —太谷核不育小麦鉴定总结。作物学报, **6** (2): 85-98。

[2] 邓景扬、高忠丽, 1982, 小麦显性雄性不育基因的发现、鉴定及在遗传学和育种学上的价值, 中国科学, B辑 (1): 45-57。

[3] 吴兆苏, 1983, 美国大麦、小麦雄性核不育的研究利用动态, 国外农学—麦类作物, (1): 1-6。

[4] 邓景扬、纪凤高, 1983, 显性雄性不育基因Tal在小麦育种上的价值与主要利用途径, 中国农业科学, (4): 6-11。

[5] 吴兆苏、沈秋泉、陆维忠、杨赞林, 1984, 小麦抗赤霉病基因库的建拓, 作物学报, **10** (2): 73-80。

[6] Sorrells, M.E. and S.E. Fritz, 1982, Application of a dominant male-sterile allele to the improvement of self-pollinated crops. *Crops. Sci.,* **22**:1033-1035.

[7] Ramage, R.T., 1977, Varietal improvement of wheat through male-sterile facilitated recurrent selection ASPAC Food and Fertilizer Technology Center. *Tech. Bull.,* **37**: 6.

[8] Hallauer, A.R., 1981, Selection and breeding methods. In Frey, K.J. (ed.) Plant Breeding II. Iowa State Univ. Press, Ames. pp. 3-55.

[9] Redden, R.J. and N.F. Jensen, 1974, Mass selection and mating systems in cereals, *Crop Sci.,,* **14**:345-350.

[10] McNeal, F.H., C.F. McGuire and M.A. Berg, 1978, Recurrent selection for grain protein content in spring wheat *Crop Sci.,* **18**:779-782.

[11] Busch, R.H. and K.Kofoid, 1982, Recurrent selection for kernel weight in spring wheat *Crop Sci.,* **22**:568-572.

[12] Avev, D.P. and H.W. Ohm *et al.,* 1982. Three cycles of simple recurrent selection for early heading in winter wheat *Crop Sci.,* **22**:908-911.

[13] Frey, K.J., 1967, Mass selection for seed width in oat populations *Euphytica,* **16**:341-349.

AS EFFECTIVE TOOL IN DISTANT HYBRIDIZATION

Zhu Hanru   Li Xiaochun

(*Zhejiang Agricultural University*)

Introduction

Distant hybridization is very significant as an important way of synthesizing new species and of improving existing varieties. But due to a great deal of obstruction to cross fertilization, hybrid embryo development and self-bred fertility of the hybrids, the percentage of successful crosses is very low. In addition, the crossing procedures are complex and pseudo-hybrids are easily produced.

In recent years, with the development of the technology for culturing immature embryos *in vitro,* with the use of hormones in the process of fertilization, and with the discovery of various bridge materials, the percentage of successful crosses of distant interspecific and intergeneric crops has been increased. As for wheat, Riley et al. determined that the chromosome of Chinese Spring possesses a crossable gene with rye. Bao Wengkui et al. called Chinese Spring the bridge type variety. Snape reported that the crossability between Chinese Spring and *Hordeum bulbosum* is controlled by the same gene as that which controls the crossability between Chinese Spring and rye[2—4, 6]. In order to make use of this property effectively, we have introduced the Taigu dominant genic sterile gene *Ta*1 into Chinese Spring through backcrossings [1]. We hoped to obtain material which has the property of both male sterility and crossability in distant hybridization in order to simplify the crossing procedure and to increase the percentage of successful crosses. We also expected that, by using F_1 hybrids between Chinese Spring (*Ta*1) and different wheat varieties as intermediary hybrids, distant hybridization might be carried out in order that the distant hybrids so obtained might possess a wider genetic background[4, 5]. In the present paper, the results of this study are further reported.

Experimental Materials and Methods

Between 1978 and 1981, using Zhemai No. 1(Ta1) as the female parent and Chinese Spring as recurrent parents, backcross transferences were made, and F_1, B_1, B_2 and B_3 seeds were obtained. Crosses using rye (*secale*) as the male parent and different transferred generations of Chinese Spring (Ta1) were made every year. Crosses of Chinese Spring, Yangmai No.1 and Yangmai No.1(Ta1) with rye were used as controls. The seed-set percentage of different combinations was calculated and the morphological characteristics of hybrids as well as their somatic chromosome number were observed. In the meantime, distant crosses of oats(*Avena fatua*), bulbous barley (*Hordeum bulbosum*, $4x = 28$), barley (*H. vulgare*), *Haynaldia villosa*, rye (*Secale*) and Aegilop (*Triticum kotschyi*) with Chinese Spring (Ta1) were carried out. The seed-set percentage of crosses was calculated in each instance.

By means of green house cultivation and illumination treatments adjustments in the time required for blooming of all parents were made. For most of the crosses, artificial pollinations were employed, made spike by spike and floret by floret. Pollination was generally done one or two times daily. After pollination, a solution of gibberallic acid at 75 ppm concentration was sprayed on seeds once daily.

Results and Discussions

Seed-set of crosses between different generations of Chinese Spring (Ta1) and rye

The results of seed-sets from F_1, B_1, B_2 and B_3 crosses obtained from transferred crosses between Zhemai No.1 (Ta1) as the female parent and Chinese Spring as the recurrent parents, with rye, are listed in Table 1. From this table, it can be seen that, compared with cross combinations between Chinese Spring, Yangmai No.1,Yangmai No.1 (Ta1) and rye, the crosses between different transferred generations and rye possesed crossability to a certain extent, Furthermore, with the advance of transferred generations, the degree of crossability was gradually increased. After three backcrosses, the crossability of Chinese Spring (Ta1) in distant hybridization had nearly attained the level of Chinese Spring. F_1 hybrids between transferred generations of Chinese Spring (Ta1) and rye possessed the characteristics of three parents, i.e. Zhemai No.1 (Ta1), Chinese Spring and rye. The PMC of F_1 hybrids crossed between Chinese Spring (Ta1) B_1 and rye was observed and the chromosome number was found to be 28. In a few cells some polyspory was found. F_1 plants were highly sterile, but in one backcross to common wheat, fertility was restored to some extent. Six plants of B_1F_3 were obtained in 1983 and all three had some characteristics of both wheat and rye. The stem, leaf and spike of one of them were similar to those of wheat. This plant had long awns and hair-glumes like rye, grew normally and grew relatively plump seeds. The chromosome numbers of root-tip cells of this plant were mainly 42 and rarely 44. whether this wheat type plant originates itself from chromosome substitution or from chromosome addition is a problem to be further analysed.

It is worth noticing that the average seed-set percentage obtained was 15.7% when the F_1 hybrids between Zhemai No.1 ($Ta1$) and Chinese Spring were crossed with rye. Therefore, it may be assumed that if F_1 hybrids between Chinese Spring ($Ta1$) and various wheat varieties, or if F_1 hybrids between various varieties ($Ta1$) and Chinese Spring, as intermediary varieties, were crossed with rye, not only wheat-rye hybrids could easily be obtained, but also these hybrids could possess various different genetic backgrounds respectively. Thereby this would provide an effective way for expansion of the wheat-rye germplasm pool. The following concrete practice may be adopted. Chinese Spring ($Ta1$) with various wheat cultivars were planted in every other row, and crosses were carried out through open-pollination. From the plants of Chinese Spring ($Ta1$), hybrids were obtained. These hybrids used as intermediary hybrids, were grown in isolated plots of various types of rye, and thereby a large number of wheat-rye hybrids with different genetic backgrounds could be obtained for selection in the breeding program.

Crossability of crosses between Chinese Spring ($Ta1$) and other cereals

Experimental results showed that varying amounts of seeds were obtained from crosses between Chinese Spring ($Ta1$) and other cereals tested; except that no seeds were obtained from crosses between interspecific oats and intergeneric *Haynaldia* (Tab. 2). The seed-set percentage of crosses with *T. kotschyi* is the highest one, equal to 45.2%, while with rye it is equal to 43.8%. When Chinese Spring ($Ta1$) was crossed with *H. bulbosum* and *H. vulgare,* a few poorly developed seeds were also obtained and their seed-set percentages were 1.2% and 0.48% respectively. The seed-set percentage of hybrids between Yangmai No.1 ($Ta1$) and *T. kotschyi* was 50.1%, whereas that between Yangmai No.1 ($Ta1$) and *secale* was a mere 1.01%. No seeds were obtained when Yangmai No.1 ($Ta1$) was crossed with other cereals tested.

Table 1. Seed-set percentage of crosses of different generations of Chinese Spring(Ta1) with rye

years	cross combination	numbers of florets crossed	numbers of seed-set	average seed-set percentage	range of fluctuation
1979	Chinese Spring(Ta1)F_1/rye	172	27	15.7	12.9–17.3
	Yangmai No.1(Ta1)/rye	82	0	0	0
1980	Chinese Spring(Ta1)B_1/rye	198	69	34.8	32.8–38.9
	Chinese Spring/rye	108	59	52.6	48.4–63.6
	Yangmai No.1/rye	76	2	2.6	2.6
1981[1]	Chinese Spring(Ta1)B_2/rye	1701[2]	102[2]	6.1[2]	
1983	Chinese Spring(Ta1)B_3/rye	720	564	78.4	69.4–81.1
	Chinese Spring/rye	74	63	85.1	68.0–89.2

1) Crosses were carried out by natural pollination.
2) All figures were recorded from observations of random samples.

Table 2. Seed-set percentage of hybrids between Chinese Spring(Ta1) and other cereals [Yangmai No.1 (Ta1) as control]

female parent \ male parent	Av. fatua			H. bulbosum			H. vulgare			Haynaldia			Secale			T. Kotschyi		
	number of florets	number of seed-set	%	number of florets	number of seed-set	%	number of florets	number of seed-set	%	number of florets	number of seed-set	%	number of florets	number of seed-set	%	number of florets	number of seed-set	%
Yangmai No.1 (Ta1)	108	0	0	3,136	0	0	46	0	0	/	/	/	394	4	1.01	96	49	50.1
Chinese Spring (Ta1)	134	0	0	4,764	570[1]	1.2[1]	3,738	17.0[1]	0.48	52	0	0	2,041	885	43.8	58	27	45.2

1) Including seeds with shrivelled and indistinct embryo.

The distant crosses between Chinese Spring (*Ta*1) and partial cereals showed crossability similar to that of Chinese Spring. This indicated that the gene *Ta*1 and the crossable gene of Chinese Spring may be expressed independently. By means of distant hybridization with Chinese Spring (*Ta*1), not only may artificial emasculation be saved and pseudo-hybrids be eliminated, but also natural pollination can be utilized in isolated plots. It is expected that the scope of crosses may be expanded and progenies of higher affinity may be obtained.

References

[1] 邓景扬等，1980，小麦显性雄性不育基因的发现与利用—太谷不育小麦鉴定总结，作物学报，6（2）：85-98。

[2] 鲍文奎等，1981，八倍体小黑麦育种与栽培，贵州人民出版社，40-44。

[3] 汪丽泉、朱汉如等，1982，中国春小麦（6X）×苏联球茎大麦（4X）属间杂交的研究初报，作物学报，8（2）：96-101。

[4] 朱汉如、李晓春等，1983，显性核不育"桥梁"小麦的转育及其不同世代与黑麦可交配性的测定，山西农业科学，（5）：20-22。

[5] 朱汉如、李晓春等，1983，显性核不育小麦衍生材料及其属间杂种在属间杂交中的利用，浙江农业大学学报，（4）：361-365。

[6] Snape. J.W. and E. Simpson (1978). Crossability of wheat with *Hordeum bulbosum*. Plant Breeding Institute. Annual Report,Cambridge, England, pp. 132-133.

Application of the *Ta*1 Gene to Triticale Breeding Program

Ji Fenggao   Deng Jingyang

(Institute of Crop Breeding and Cultivation, Chinese Academy of Agricultural Sciences)

Introduction

The triticale breeding includes two major objectives: 1) to synthesize primary triticale strains for enriching its germplasm; 2) to improve the synthesized triticale strains, to composite more positive traits into one strain and then to make the constant improvement of genetic background needed by the major genes, in order to breed varieties which can compete with wheat in production. *Ta*1 gene may play a role in these two objectives.

Creating more primary triticale strains by means of *Ta*1 gene

Making crosses between wheat and rye and doubling their chromosome numbers is the central task for the creation of the primary triticale and the enrichment of its germplasm. But the first obstacle facing us, a difficult one, is to get a large number of true hybrid seeds. Only some wheat varieties can be easily crossed with rye but very few varieties crossed with rye will have as high fertility as that of wheat intraspecific crosses. Lein(1943) thought the crossability between common wheat and rye is controlled by 2 recessive genes kr_1 and kr_2. Riley and Chapman (1967) localized Kr_1 and Kr_2, which have the opposite effect of kr_1 and kr_2, on the chromosome 5B and 5A, respectively. Bao Wenkui utilized the varieties having better crossability as the bridge varieties, to be crossed with the poor crossability varieties and then crossed F_1 or F_2 with rye to get large number of primary trilicale strains[1]. However, because the techniques required, like hand emasculation and pollination, are timeconsuming and delicate, there will be inevitable faults in the operation. These cause selfing seeds or intraspecific hybrids, yet eliminating these false hybrids is not only rather troublesome but also quite necessary. Making crosses of male sterile Taigu wheat with rye has the advantage here as no emasculation is needed; hybrids are always true and reliable.

Making use of $Ta1$ gene together with kr_1 and kr_2 genes

The sterile plants from the open-pollinated population has low fertility when crossed with rye. A population we measured only set 2.7 kernels per head in average, with great variation between plants. This is due to the low frequency of kr_1 and kr_2 genes in the random mating population. Six female parents were F_1 hybrids from the crosses of open-pollinated male sterile Taigu wheat with six spring wheat varieties: PH88, (Zhong 7605/PH88) F_5, Zhong 7725, Zhong 7901, Penjamo 62 and Yecora. At the same time, the F_1 hybrids were crossed and backcrossed with Chinese Spring for several times to get three-way cross F_1, BC_1, BC_2 seeds, etc. All these F_1, BC_1 and BC_2 male sterile plants were crossed with 4 rye varieties: Jingzhou rye, Snoopy-2, Snoopy-3 and Lanzhou rye. The average seed setting from the crosses of female parent F_1, BC_1 and BC_2 with rye are listed in Table 1.

There were great differences of seed setting between plants from the crosses of (Taigu wheat/Chinese Spring) F_1 sterile plants with rye. The maximum measured was up to 235.3 kernels/100 spikelets, while 23.1% of plants gave no seeds at all. The maximum seed setting of BC_1/rye was 206.9 kernels/100 spikelets, while no non-seeding plants were found. The difference between high seed setting of F_1 and BC_1 plants might be due to the diverse environmental conditions. The different crossability between plants within the same generation could be caused by the random distribution of kr_1 and kr_2 genes in the population. As the increasing of frequency of these genes and number of backcrossing times, the average seed setting also increased (Plate XIII, 1–2). The results investigated on the early backcross generations are listed in Table 2.

Except for the investigated plants, we made the selections on the BC_1 plants for crossability. The main heads of sterile plants were crossed with rye. The tillers of the plants having high seed setting were crossed continuously with Chinese Spring, those plants having low seed setting were pulled out. In some plants which were earlier to flower than rye, their main heads were used to backcross with Chinese Spring but their tillers were crossed with rye. The hybrid seeds were accepted or rejected depending on their crossability with rye. This method was proved effective in practice. The crossability of BC_2 plants was increased to 229.2 kernels/100 spikelets. When

Table 1. Seed setting per 100 spikelets of 6 female parents and their cross and backcross progenies, with Chinese Spring (C.S.), crossed with male parent rye

combinations female parents (F.P.)	F.P./rye	F.P.₁/C.S.//rye	F.P./2x C.S.//rye	F.P./3x C.S.//rye
Taigu wheat/PH88	9.47	45.84	105.3	274.6
Taigu wheat//(Zhong 7605/PH88) F₅	5.11	31.75	81.3	214.3
Taigu wheat/Zhong 7725	5.53	48.74	95.4	217.2
Taigu wheat/Zhong 7901	6.44	57.43	85.6	321.1
Taigu wheat/Penjamo 62	2.07	24.33	54.7	215.0
Taigu wheat/Yecora	3.33	28.54	67.7	132.6
average	5.33	39.44	81.7	229.2

Table 2. The increase of crossability along with the increased number of crossing times with Chinese Spring¹⁾

No. of plants Crossing times with C.S.	1	2	3	4	5	6
F₁/rye	0	0	23.8	87.6	153.7	235.3
BC₁/rye	18.6	56.7	92.7	123.2	191.5	206.9
BC₂/rye	87.3	122.2	199.4	251.0	316.3	287.6

1) The numbers in the table were kernels/100 spikelets.

2 basal florets were investigated, the mean fertility was 86.4%, but hand emasculated Chinese Spring/rye was 85.1% fertility, both of which were rather close. The former fertility figure may be a little high not having any damage brought by emasculation.

The above results, like that from Zhu Hanru[2], indicated that $Ta1$ gene has no interaction with kr_1 and kr_2 genes, so they can be integrated into one plant to create the bridge varieties or population containing $Ta1$ gene. We have transfered $Ta1$ gene into some varieties like Jiangtongmen, Mazhame, Beyiuzhi, etc, which have good crossability with rye. In transferring $Ta1$ gene, the male sterile Chinese Spring is one of the best $Ta1$ gene donors. Thus kr_1 and kr_2 gene can be brought into the acceptor at the same time. Other wise, the crossability with rye will decrease.

Producing wheat/rye hybrid by free pollination

In the spring sowing of 1983, we mixed the same amount of $6BC_1$ line seeds and $6BC_2$ line seeds as two female parents and 4 rye variety seeds as the male parent. Two rows of male and female parents were alternatively planted in the isolated plot. Before anthesis, all the fertile plants in the female parent rows were eliminated by daily checking and sterile plants were labelled. 208 heads and 1678 hybrid seeds from BC_1 female parents were harvested with the average fertility of 8.07 kernels/head. 247 heads and 2853 hybrid seeds from BC_2 female parent rows were harvested and fertility was 11.55 kernels/head in average. In the $30m^2$ plot, only 5 work days were spent to get 4531 hybrid seeds.

The experiment in 1983 was a small scale. This year, we have expanded the experiment, from spring wheat to winter wheat. The female parents included the free pollinating population and F_1 hybrid between the bridge varieties with the superior varieties.

The traits of F_1 hybrids

All F_1 hybrids between Taigu wheat and rye, in any of crosses, can be clearly distinguished at flowering into two types: small anther group with $Ta1$ gene and large anther group without $Ta1$ gene, having ratio of 1:1 (Plate XIII 3-4).

In the group without $Ta1$ gene, anthers change colour from milk-white to light green at meiosis, appear yellow green prior to anthering, and look yellow at flowering stage. At full bloom, yellow anthers are hanging all over the heads but anthers are slender and unable to dehisce (Plate XIII, 5). By cytological study of PMCs, one can see the random distribution of 28 chromosomes, uneven tetrads in different size at quartets and all kinds of abnormal pollen grains at flowering stage. Some pollen grains have a little inner content, some are empty and only a few pollen can become blue in I-KI solution accounting for 0.2–0.4% of the total. Because their anthers fail to dehisce, almost all plants in this group can not give seeds without foreign pollen grains. After chromosome doubling, part of anthers from some plants can develop well and dehisce. Most of these plants can set seeds in different number by selfing, so as to become the new primary triticale strains.

In the group carrying $Ta1$ gene, anthers stop developing from meiosis to monokaryotic stage and remain milk-white. The stamen abortion happens at the same stage and to the same extent as male sterile Taigu wheat (Plate XIII, 6). The pistils look normal but very few plants can set seeds by pollination of octoploid triticale. These plants, after being treated with colchicine at tillering stage, become short in height with their pistils being similar in appearance to those of nontreated plants. In the first treatment, we found 17 small anther plants, 15 of which set seeds in different numbers by being pollinated with the octoploid triticale.

Although this approach has some advantages, there is also a disadvantage, i.e. only half F_1 hybrids can be directly used to produce primary triticale strains; the other half plants carrying $Ta1$ gene can only be used as parents because they are male sterile.

The development of the dominant male sterile octoploid triticale and their utilization

More than 100 years have passed since Wilson developed the first hybrids between wheat and rye in 1875. However, it was very difficult to obtain a new primary triticale strain before the discovery of chromosome doubling by colchicine treatment. There was a upsurge in the wide hybridization after 1936 and the triticale breeders developed many new primary triticale strains. Although many of these strains have good traits including better disease resistance, strong environmental stress tolerance, wide adoptability, big heads and more florets and high content of protein and lysine, almost all of them have defects such as low fertility, poor plumpness, late maturity and excessive height. Thanks to the efforts of breeders, such characteristics as high fertility, good plumpness, short stature and early maturity have been developed in the triticale; nevertheless, they have not been combined in one strain[1, 5]. Therefore, to combine the triticales with different good traits has become the goal in triticale breeding programs. Comparing with the widely cultivated crops, triticale has never experienced long term natural selection. Artificial selection needs to be carried out on a large scale for many years.

Large numbers of gene recombinants and higher selection pressure could be quite necessary for integrating the good genes and improving their genetic backgroud because the genetic components from wheat and rye have not yet been highly equilibrated and harmonized in this synthetic species. Now, it is the dominant male sterile gene that is the desirable tool for gene recombinant of self pollinated crops. Sorrells and Fritz[8], Deng and Ji[3], Zhang and Ji[4] planned the programs in application of the male sterile dominant gene to the recurrent selection in the self-pollinated plants, the program is also applicable to triticale. The key point is to develop triticale strains which carry the dominant male sterile gene.

We found that $Ta1$ gene is rather stable in further study of the male sterile dominant Taigu wheat. $Ta1$ gene can be expressed normally in the F_1 hybrids between the male sterile Taigu wheat and rye. $Ta1$ gene can also be expressed stably in the amphiploids from the chromosome doubling treatment. These male sterile amphiploids were pollinated with the existing octoploid triticale to produce the F_1 offspring, which were all male sterile because $Ta1$ gene is dominant. As these male sterile plants were again pollinated with octoploid triticales, the next offspring had segregation: about half fertile and half sterile (Plate XIV, 1-2). The fertile plants were similar to the F_1 hybrids between octoploid triticale strains, which were self-fertile, but the fertility varied greatly among individual plants with the range from 0 to 90%. Seeds from the fertile plants grew into plants, all of which had large anthers and no stamen abortion at all (Plate XIV, 4). The sterile plants had very weak and small anthers, about half the length of the large anthers. Most of them stopped developing from meiosis to late monokaryotic stage and very few anthers could be seen hanging outside of glumes at full blooming stage. On the contrary, the pistils were well developed with glume opened and stigma protruded (Plate XIV, 4). These sterile plants can set seeds when pollinated with all kinds of triticale strains and their offsprings once again show half sterile and half fertile segregation. Thus, extensive continuous gene recombinants can be conducted by crossing with the sterile plants. While using the different parents with desirable traits, the recurrent selections and recombinations are carried out to increase both the types of combination and the frequency of superior genes in the population so as to produce better recombinants. At the same time, from fertile plants segregated from each generation, new triticale strains or varieties having still better traits could be released through careful selection. At present, we have began to transfer $Ta1$ gene into 126 triticale strains or varieties which have distinctive characteristics and we will start the triticale recurrent selection after further raising the proportion of the paternal genetic component in the population.

After the chromosome doubling of F_1 hybrids of male sterile Taigu wheat/rye, the plants carrying $Ta1$ gene could also give seeds when pollinated with hexaploid triticale, and its fertility, seed plumpness and germinating ability were all good. F_1 hybrids mostly show male sterile and these sterile plants were able to set seeds when pollinated with octoploid or hexaploid triticale, respectively. However, the seeds from former pollination were partially shrivelled and unable to germinate while seeds from the latter pollination mostly were plump and viable. Crosses were made between newly-bred male sterile octoploid triticale and hexaploid triticale and F_1 hybrids were backcrossed with both parents. The results were similar to the above stated one, except for the F_1 segregation in fertility. In short, $Ta1$ gene can be employed in the gene flow between octoploid and hexaploid triticale. It is easier to transfer a good characteristic from octoploid triticale to hexaploid one than the reverse.

Liu Binghua, et al., have located $Ta1$ gene on the short arm of 4D chromosome. It is difficult to create male sterile hexaploid triticale unless the addition, substitution or translocation lines involving 4D chromosome are made. If a translocation line carrying $Ta1$ gene is developed, hexaploid triticale breeding could be pushed forward.

We have done other experiments in which the dominant male sterile octoploid triticale can also be developed by crossing the male sterile Taigu wheat with the triticale as the recurrent male parent. It is very easy to cross male sterile Taigu wheat with triticale, but resultant seeds will not germinate and grow into plants. About two thirds of the seeds had anomalous endosperm and embryo which could not develop into seedlings by embryo culture. The other one third of the seeds had no endosperm at all. Part of them had well-developed young embryo, which could quickly germinate and grow into seedlings on the artificial media (N_6 or MS media). F_1 hybrids obtained from embryo culture had flourishing vegetative growth and vigourous plants, which could be distinguished into two types: sterile and fertile. In fertile plants, the anthers can dehisce and shed pollen, but very few plants set seeds by selfing. The sterile plants can give seeds by pollination of octoploid triticale, but a number of these seeds were abnormal and another embryo culture is needed to get more backcrossing progenies.

In the experiment of crossing the male sterile Taigu wheat with hexaploid triticale, the situation was the same but sterile plants in the backcross were not so many as fertile plants.

In the cross of male sterile Taigu wheat with the octoploid *Agropyron-triticum*, the hybrid seeds were not so plump but they could germinate into seedlings. When backcrossing with the octoploid *Agropyron-triticum* F_1 hybrids had about 50% fertility. With this method, we have obtained the dominant male sterile octoploid *Agropyron-triticum*.

The dominant male sterile *Ta*1 gene can be stably expressed in the haploid and amphiploid of common wheat/rye, in the heptaploid of octoploid triticale/hexaploid triticale, in the progenies of common wheat/octoploid or hexaploid triticale and in the common wheat genetic background containing part of *Agropyron* genetic materials. It seems that *Ta*1 gene can be extensively used with multiple approaches in the wide hybridization and chromosome engineering in which common wheat is one of the leading parents.

References

[1] 鲍文奎主编，八倍体小黑麦育种与栽培，贵州人民出版社，1981年。
[2] 朱汉如等，1983，显性核不育"桥梁"小麦的转育及其不同世代与黑麦可交配性的测定。山西农业科学，(5): 20-23。
[3] 邓景扬、纪凤高，1983,显性雄性不育基因Tal在小麦育种上的价值与主要利用途径，中国农业科学，(4): 6-11。
[4] 张绍南、纪凤高，1984，利用显性雄性不育基因进行小麦轮回选择育种的选择与交配问题，作物学报，10 (3): 154。
[5] Triticale: First Man-made Cereal, 1974, A Symposium of 58th (1973) annual meeting of the American Association of Cereal Chemists (AACC).
[6] Triticale, 1973, Proceedings of an international Symposium. E1 Batan, Mexico.
[7] Müntzing, A., 1979, Triticale Results and Problems. *Suppl. to Jour. of Plant Breed.* No. 10.
[8] Sorrells, M.E. and S.E. Fritz, 1982, Application of A Dominant Male-sterile Allele to Improvement of Self-pollinated Crops. *Crop Science,* 22(5): 1033-1035.

Plate XIII

1. The spike of fertile plants from male-sterile Chinese Spring.
2. The spike of sterile plants from male-sterile Chinese Spring.
3. The spike of large anther plants of F_1 hybrid of male-sterile Taigu wheat × rye.
4. The spike of small anther plants of F_1 hybrid of male-sterile Taigu wheat × rye.
5. Pistil and stamen of large anther haploid plant F_1.
6. Pistil and stamen of small anther haploid plant F_1.

Plate XIV

1. Spike of fertile plant from male-sterile octoploid triticale.
2. Spike of sterile plant from male-sterile octoploid triticale.
3. Pistil and stamen of fertile plant from male-sterile octoploid triticale.
4. Pistil and stamen of sterile plant from male-sterile octoploid triticale.